Vulture Culture 101: A Book For People Who Like Dead Things

For my fellow Vulture!

By Lupa

23 Nov '22

2019

http://www.thegreenwolf.com

Vulture Culture 101: A Book For People Who Like Dead Things
Lupa
© 2019 First edition

Text, editing, cover art/design and layout by Lupa
Guest essays by Divya Anantharaman, Ashley Cheeks, Eric Foote, Shelby Hendershot, Escher Null, and Amy Wilkinson.

Set in Garamond and VintageOne by Jake Luedecke
Self-published by Lupa
http://www.thegreenwolf.com

ISBN-13: 9781092885102

Dear readers, I appreciate that books can be expensive, and free ebooks may be the only resource some of you are able to access. I know it's frustrating to not always be able to afford the resources you want, but I respectfully ask that if you are reading this book in ebook form that you not put copies online. Just as you need money to pay your bills and put food on the table, so I need to be able to do the same for myself and my family. Please respect that by paying for this book rather than using pirated copies, so that I can continue to afford to invest the time it takes me to write and produce books for you to enjoy and learn from.

If you'd like to read some of my material for free, my blog is at http://www.thegreenwolf.com/blog. You are always welcome to ask me questions, too; it may take me a few days to get back to you if I'm particularly busy, but I do try to answer every email and message I get. My email address is lupa.greenwolf@gmail.com. Thank you for your respect and consideration, and I look forward to hearing from you.

Disclaimer

The contents of this book may be considered indelicate, insensitive, or thoroughly offensive to some readers. In the event you didn't read the back of the cover or see the dead animals on the front, please be warned that the following pages extensively discuss the remains of deceased creatures, to include photos and sometimes graphic descriptions of things like skinning and taxidermy. If that sounds like it's right down your alley, read on! If not, consider giving it a try anyway; you might still find it curious from a purely theoretical perspective.

Acknowledgements

Thank you to my amazing guest essayists, Divya Anantharaman, Ashley Cheeks, Eric Foote, Shelby Hendershot, Escher Null, and Amy Wilkinson for contributing your expertise and ideas, as well as your writing and photos, to this project. A huge thanks to all of my IndieGoGo contributors, without whom this project wouldn't have happened financially. Much love to the Vulture Culture, for showing me I'm not alone in my weirdness and love of nature live and dead.

Dedication

To S., for still supporting me and my creativity for so many years, and continuing to show me love, respect, and ridiculousness. And to Pikka, for reminding me just why it is I need a dog in my life.

And a Huge Thank You To...

All of my IndieGoGo supporters! In spring of 2018, these fine folks were willing to fund the book up front through a crowdfunding campaign I ran; not only did we meet the goal but we surpassed it. My gratitude to the following:

Tasha King, Elise Wolf, Ash Windler, Stevie Miller, Damien Hunter, Juli Beighley, Liam Kopolow, Joy Anne Phillip, Joshua Mulcahy, Becki Decker, Stacey Gregorash, Jessica Hoone, Ashton John Lanning, Jeremy Taylor, Cayley Rydzinski, Jonathan Miller, Sam Rich, Genel Gronkowski, Morrigan Proude, Kaitlyn Marie Gerber, Miranda L. Tarrow, Nora Timmerman, Elizabeth Akins, Noelle French, Teri AnpoWi Saveliff, Sarah Kamins, Kathleen Eubanks, Melinda Kral, Victoria Davis, Ember Cooke, Kelcie Chase, Shannon Legler, Spencer Hughes, Ella Pressler, Kristen Fogarty, Thomas Kerr, Laura Hiner, Arwen Lynch Poe, April Broughton, Brian Ledbetter, Dawn Corley, Liv Rainey-Smith, Carol Hutte, Teunis Peters, Chelsea Mascari, Emily Siskin, Michelle Campbell, April Molohon, Aidan Bork, Elizabeth Koprucki, Gwendolynn Amsbury, Monica J. Van Steenberg, Brooke Hambright, Kate Lipford, Juliana Russell, Luke Ries, Ashley Best, Elizabeth Vattes, Tauna Bollinger, Amanda Vargas, Steve Dean, Tamra Perry, Hana Russell, Elias Lindroos, Taylor Thompson, Kendall Deutsch, Katie Crowley-Dean, Janice May Bignell, Tanya Hindmarsh, Meghan Farthing, Megan Fox, Sarah Schowengerdt, Natalee Dula, Jessica Sallay-Carrington, Jessica Marshall, Emma-Jayne Easton, Christine Delillo, Kristina Brede, Lindsay Fenwick, Ashlyn Thigpen, Rebecca Jasulevicz, Fall Thompson, Emily Harris, Laurel Pattee, Jesse Spacco, Darien Baysinger, Lauren Pavlik, Miika Ahjopalo, Sierra Custer, Brannon Usko, Karla Elms, Carolyn Neal-Harden, Dubout Erwan, Blake Bethea, Sam Moran, Marnie Gordon, Marie Alessandro, Brenda Lively, Larry Andrews, Alice Cartes, Abigail Allen, Tamara Siuda, Cheryl Kirk, Stephanie Baldridge, Shelly Ferland, Alexa Swenson, Kit Ewing, Brynn Maurer, Amy McCune, Danielle Cromer, Jennifer Page, Hannah Wallenbrock, Amanda C. Mills, Carrie Shane, Percy Delacroix, Lisa Olivas, Kalen M Boley, Jack Vivace, Lae, Brooke Hambright, Rachel Greenfield, Amanda Nixon, Brooke Hambright, Gryffin Mutzenberger, Jess Draws, Darzie, Amber Stewart, Galeogirl, Michelle Crawford-Bewley, Jasper Jacobs

Table of Contents

Introduction

It is an undisputed fact that a child can imbue the most mundane item with immeasurable value through the ancient tradition of "finders, keepers". It is also a supreme tragedy of childhood that, all too often, some well-meaning adult discovers this unlabeled treasure, mistakes it for garbage, and sends it on a one-way trip to the landfill. I lost an entire hoard, cached in an old dusty shoebox, in exactly this manner.

I grew up in the 1980s in a small Missouri town generously pockmarked with patches of woods and fields. My parents were of the opinion that since they'd been given free rein over their childhood neighborhoods, I should have the same right to roam. As long as I stayed within earshot of the ancient cowbell my mother would clang when it was time to come home, I could wander through neighbors' yards and open lots at will.

This was the birth of my love affair with all things outdoors. And I do mean *things* in a literal, physical sense. From an early age I arrived home with my pockets toting all manner of little treasures. Many were benign by most standards: flowers, shiny rocks, pine cones, old glass bottles frosted by years of grit and storms. I was allowed to keep live grasshoppers, box turtles and garter snakes on the porch so long as they didn't come inside and I let them go after a couple of days.

I assigned value to my various finds according to how easily obtained they were. Flowering crown vetch could be gathered by the handfuls from the slope along the retirement home; limestone outcroppings made for easy pickings if I needed play building materials. These were beloved, but easy to come by. The turtles and snakes were more coveted, and worth the risk they posed to tender young fingers. Cottontail rabbits were all but unattainable, too swift for my clumsy, gangling legs.

Rabbit *bones* were another matter entirely. Digging through the vetch, I would come across a sun-bleached skeleton every so often, and the fist-sized skulls were among my most highly prized finds. I carefully extricated them from their half-burial in the dirt and greenery, and carried them home where I would secret them away in my shoebox in the garage. There I could admire them in detail, along with a single turkey wing feather, a piece of galena, a large chunk of raw iron ore.

It wasn't just their aesthetic appeal that made me keep them. They symbolized the freedom I sought from the constrictions of classrooms and Sunday Mass; from the pale glow of sunrise, through the long summer afternoon, and into the constellations of fireflies, they waited for me in my shoebox of treasures. All I had to do was feel the hard ridges and brittle planes, and even if I had to return to school the next day, I knew that my wild places—and my wild self—were waiting for me when I was done.

Late in 1990, my family moved to a different house across town. Unfortunately somewhere in the process my little box of treasures ended up lost—

perhaps thrown out by mistake, or deliberately removed because the contents were unsettling. Not a great tragedy in the grand scheme of things, but devastating at a time when I was being uprooted from everything I held dear. While I quickly set about exploring my new neighborhood, I might have felt less disconnection from my old stomping grounds if my treasure box had made it safely with me. Of all the items that were lost, I missed those cottontail skulls most of all; unlike decaying leaves and real estate signs, they were constant and unchanging, something I could have used in my turbulent teen years, when I clung to fading threads of the wild places that had raised me.

Most people in the United States (and in plenty of other places) are worryingly detached from non-human nature. We spend the bulk of our time indoors; "outside" is the space we traverse to get to the next building. From the time we're born in a hospital maternity ward, to spending six to eight hours a day in a school building, to repeating that process in the workplace, few of us have more than a cursory amount of time each day without a roof over our heads.

But we yearn for the wide, open spaces that our ancestors evolved in over millions of years. Have you ever been in an office building whose windows look out over a parking lot but whose walls are covered in large photo prints of waterfalls? Or visited a house where the décor trends heavily toward fern stencils, glass bowls of shells and pine cones? These are all trappings of a wilderness we've long since abandoned as a culture, yet need on a very instinctive level.

As children, we are closer to our mammalian heritage; our tendency to pick up oddments off the ground reflects our natural curiosity. It's only as we get older that we're discouraged from such unsanitary practices—unless we're scientists doing field work, or zookeepers, or otherwise engaged in work where the outdoors is unavoidable. If you're over the age of, say, twelve, and you're still bringing home bones and feathers without a socially acceptable work-related excuse, you may find yourself on the receiving end of the stinkeye.

Yet there are those of us whose fascination with natural specimens is more powerful than social pressures. In an environment that is so wholly human-dominated, we want to surround ourselves with reminders of non-human nature. We *need* to be able to sense the world beyond our anthropocentric echo chamber. We're losing our wild selves at the same rate we're losing wilderness: we've traded nature wisdom for artificial comfort.

I share that experience with many others within Vulture Culture. Born in 1978, I am of the last generation where television was the primary factor distracting children from embarking on outdoor explorations. It is no surprise to me that so many self-proclaimed Vultures are in their teens and early twenties now, raised on not just T.V., but computers and video games, too. Our shared desire to not lose sight of nature binds us together more tightly even as technology increases its powers of diversion.

Our collections are our conduits to non-human nature, though the manner in which we connect varies from person to person. For me, it's the comparative anatomy of my skull collection that grounds me in this world. I can run my fingers over the curve of a badger mandible, then trace the same angle along my own jaw. We both have pointy canines and snipping incisors, and though our molars went in different directions evolutionarily speaking, the basic function remains the same. Despite the insistence of religion and human hubris, I am just another mammal, one among many species. The skulls, in all their variations on a common theme, keep me humble.

Not everyone has quite so complicated a connection with their specimens. For some Vultures, it boils down to "Oooooh, this looks really, really cool! I MUST HAVE IT!" Others include hides, bones and other remains in a wider aesthetic collection that may run the gamut from vintage medicine bottles and photographs to avant-garde fine art hidden in thrift store frames. All of these, and more, are valid reasons for Vulture-dom on an individual basis.

Even as you explore your motivations for bringing home birds' nests and seashells, I'd like you to also look through the wider lens of the previously mentioned yearning for nature, lost and found. Not only will it give you more context for the rest of this book, but it might even inspire further evolution in your relationships with the specimens you hold, and the places they came from.

For now, let's get some background on Vulture Culture itself.

Chapter 1: Roots and Revival: Historical and Modern-Day Vultures

So what's "Vulture Culture", anyway? Well, it's a 1984 album from The Alan Parsons Project. It's Eric Gerst's book on the predatory nature of the insurance industry, and the name of a German punk band. And, starting with Tumblr user Lifebender, it's been applied to those of us who enjoy collecting, processing and making art with hides, bones and other animal remains[1].

For several decades, taxidermy has increasingly been considered the domain of redneck uncles with stuffed deer heads on cigarette-stained panel walls. The only other people who could get away with having dead birds in the freezer were eccentric field biologists, who could also be forgiven for squirreling away owl pellets with their tiny rodent-bone fragments snuggled up against an old butter container of unmarked animal poo. Children have always brought home natural treasures, but after a certain age Mom and Dad's opinion on the matter degenerates from being merely "that's a little gross" to "why can't you just shoplift lip gloss like other sixteen-year-olds?"

That began to change in a big way about a decade or so ago, when hipsters began in earnest to make PBR and trucker hats ironically cool. Dusty old taxidermy suddenly emerged from the corners of attics and backyard sheds, relocating to mantles in communal houses and observation decks in pseudo-dive bars. True, those mounts with antlers often ended up repurposed as hat racks. But they now had social cachet.

This was good news for those of us who had kept our Vulture-dom under the radar. Taxidermy was no longer gross and creepy: it was downright trendy. While hipsters' investment in natural history specimens was a surface aesthetic that could just have easily been swapped out for velvet clown paintings or Elvira's beer advertisements, a quieter community was forming, fueled by a deep, abiding love of nature—and the internet.

But contrary to the bemused writings of alt-culture newspaper writers, we Vultures are just the newest in a long line of collectors. Royals, nobles and other wealthy folk have long displayed both living and preserved animals and plants in their homes. Often this was just a way to say "I have so much money and power that I can have someone bring me something *really weird* from all the way around the world just because I said so!" But there were genuine naturalists in the mix, too, and there's one particular breed of them that I'd like to explore in a little more detail.

The Cabinet of Curiosities

There's a common misconception that after the fall of the Roman Empire Europe descended into utter squalor and chaos, popularly known as the Dark Ages. While

---

1 http://lifebender.tumblr.com/post/63306554883/a-name-for-the-bone-collecting-community

certain elements of the economic prosperity, health and overall well-being of Europeans took a downward shift during this time, the land was not completely lacking in intellectual resources. In spite of the Catholic Church's periodic attempts to stamp out anything that challenged its power, philosophers and other thinkers were busy trading and debating ideas. And other areas like the Middle East were experiencing great periods of knowledge and prosperity. Europeans may have been having a bit of a tough time, but they weren't entirely cut off from the world and its evolving ideas.

The Italian Renaissance that began in the 14th century signaled a shift in cultural attitudes toward secular knowledge and the power of the Church, and renewed interest in the sciences and the visual arts. Increased global trade further encouraged European curiosity, particularly as overland roads were joined by an ever-growing number of ocean trade routes. As European traders and explorers headed to new (to them, anyway) lands, they discovered all sorts of exotic animals, plants and other natural phenomena, and decided to bring them back home. While live specimens were certainly prized and carried back with the greatest care possible, it was much easier to transport those that were already dead, particularly in the case of animals.

Once back home in Europe these collectors would display their new finds in special rooms. These early *wunderkammern*, or "wonder cabinets", weren't always what we would recognize today as scientifically accurate. Reality mixed all too easily with mythology even as far back as the Roman Empire, and a popular text on all things scientific was Pliny the Elder's *Natural History*, published in 79 A.D. While not entirely without merit, the book placed dog-headed men and dragons next to elephants and hyenas. If a curiosity collector happened to receive a narwhal tusk, there was a very good chance it would get labeled in all seriousness as a "unicorn horn".

Thankfully, along with a passion for acquiring neat stuff these collectors also had a love for labels. They began sorting their specimens by appearance, region they were collected from and other physical details. And because there was often a good-natured competition to see who could get the weirdest specimens that nobody else had in *their* collections, after a while everyone knew what everyone else had. So notes and observations were traded, and there began to be more of a consensus of what beings were related to each other and how they were best categorized. By the time Carl Linnaeus began creating a formal taxonomic system in the Enlightenment period of the mid-18th century, European naturalists had already largely ousted the mythical beings from their science, and *wunderkammern* became more about the very real wonders populating an increasingly global landscape.

One of the major problems with the cabinet of curiosities was that, for the most part, it was a privately accessed collection reserved for the wealthy and powerful. In later years more everyday people would have their own smaller collections, but even these required a certain amount of space, time and money to maintain. It would take a much more democratic concept to make the natural sciences accessible to all: the public museum.

Museums and Their Specimens

Imagine you are a member of a mid-19th century noble family in England. You've decked out your ancestral home in the latest styles, your gardens are immaculate, and you're looking forward to unveiling the new renovations at your next social event. Except there's just that one room, that one problematic leftover from a long-dead progenitor. It's full of animal carcasses, dried bits of plant matter, and more rocks than you know what to do with. Family legend has it that some of these items are over two hundred years old, but you couldn't care less. Still, it would cause a fuss in the family if you just threw it out. What to do?

Well, you're not all that far from the University of Oxford; maybe some of the strange old men wandering the halls would find these things amusing. You write up a letter offering to make a donation of the entire collection to the school, if only they would send someone to pick it up themselves. With a few quick pen strokes, you've both offloaded a proverbial albatross, and made yourself a benefactor of the University.

This isn't such a far-fetched story. Oxford's Ashmolean Museum was founded in 1683, built on the donated cabinet once owned by Elias Ashmole. The museum was opened to the public soon after it was completed, and became one of the earliest examples of public natural history exhibitions. The late 18th century saw an increase in the number of museums in Europe, a trend that continued all the way through the 19th century and spread worldwide. In the United States, it was billionaires like Andrew Carnegie, rather than ancient noble families, that funded the construction of museums to be populated with a dizzying array of natural history specimens, historical artifacts and other treasures for the enjoyment of the people.

Now anyone who could pay the admission fee could enjoy the wealth collected from all over the world. Throughout the 20th century museums became an increasingly popular destination for class field trips and weekend family outings. They sponsored research, to include in remote areas not studied before. They offered resources to undergraduate and graduate students alike who came to study the specimens both on display and behind the scenes. And, more sadly, they sometimes housed the last remains of animals that had gone extinct.

As scientific knowledge about wildlife increased, the displays of taxidermy, skeletons and other collections improved. Mounts were created in more realistic poses; skeleton articulations were given makeovers so they were more true to life. Environmental preservation became a greater focus in the museums' exhibits, so visitors learned not only what each animal ate and how it behaved, but what threatened to destroy it and its fragile home. The cabinet of private wonders had given birth to a wealth of publicly accessible knowledge.

Not every museum was a polished urban edifice built by wealthy philanthropists; Aunt Aggie's Boneyard typifies the smaller roadside collection of oddities created with perhaps less formal training, but almost certainly more heart. Nearly forgotten by everyone outside of a little patch of rural Florida, Agatha Jones was a pioneering artist and entrepreneur who built a miniature empire out of

Lake City, Fla.   The Boneyard, Middle Aisle

thousands of animal bones. Affectionately known as Aunt Aggie, she was an emancipated African-American slave who in 1883 bought her own property in Lake City. In addition to cultivating an exotic collection of roses, abundant fruit trees and a wild selection of native flora, she populated her land with incredible bone sculptures. The on-site natural history museum included preserved reptile specimens and other oddities on display. If there was ever a proper great-grandmama of Vulture Culture, Aggie'd be it.

By 1900, Aunt Aggie's Boneyard was a popular tourist destination. Young lovers came here to spend time together, early shutterbugs snapped pictures of the carefully crafted pathways, and some lucky folks even got their fortunes told by the grand artist herself. On the way out, visitors were presented with bouquets from the garden's many flowering plants, which no doubt increased their likelihood of leaving a tip to make sure the Boneyard would still be there next time around.

Sadly, Aunt Aggie's Boneyard was destroyed as interstates and other development encroached on the quiet little town. Agatha herself passed away in 1918, and less than a decade later a high school stood where bones had once sprouted out of the ground like ghostly saplings. Only a few pictures remain of her singular creation. But the spirit of the Boneyard lives on in every personal collection of terrariums and taxidermy, specimens collected and encouraged by those of us similarly inspired by the natural world.

The Art of Taxidermy

One of the key features of any natural history museum is its taxidermy. While humans worldwide have been preserving hides through tanning since prehistoric times, the art of arranging a hide on a lifelike form is, as far as we know, a relatively recent phenomenon. The earliest example may be stuffed, preserved birds that 14th century hunters would use to train their falcons, though these mounts probably didn't have very long lifespans in the claws of an enthusiastic raptor.

Other contemporary attempts at taxidermy generally involved a skin that was often dried rather than tanned, then stretched over a wooden form with little care for anatomical correctness. Rags may have been used to flesh it out a bit, but that was about it. Arsenic soap, developed late in the 1800s, revolutionized the art, as it both preserved the hide and kept it safe from insects and mold. There was an increasing emphasis on making the taxidermy look as lifelike as possible, and with the dawn of photography taxidermists had even more references to use than before. They carved their own forms based on the physical anatomy of the animals, sometimes incorporating the actual skulls for more accurate facial structure.

Agatha Jones was not the only person of color who is an unsung ancestor of Vulture Culture. John Edmonstone was born a slave on a Guyanan plantation in the late 1770s. Naturalist Charles Waterton took Edmonstone with him on expeditions and taught him taxidermy to preserve specimens they found along the way. In 1807 he was brought to Edinburgh and as Scotland forbade slavery this was a move to freedom. His great skill with taxidermy earned him a position at the Natural History Museum. In addition, Edinburgh University employed him as an instructor of taxidermy.

One of his students was none other than Charles Darwin. In 1825 he began teaching the disaffected medical student how to preserve and mount animal skins. Edmonstone also shared stories of his scientific discoveries while traveling with Waterton, which sparked Darwin's in natural history all the more. Darwin took both the taxidermy skills he learned from Edmonstone as well as the expansion of his world with him when he embarked on the S.S. Beagle for what would be a life-changing trip that included the Galapagos Islands. Without Edmonstone's influence, Darwin may have never gone the way of the naturalist to begin with, nor been able to preserve the specimens he collected.

By the Victorian era taxidermy had become a fine art in and of itself. Well into the 20th century particularly talented taxidermists earned names for themselves on both sides of the Atlantic, with Carl Akeley, John Hancock and Martha Ann Maxwell being just a few of the best-known examples of the time. While stuffed mice in tiny suits may seem like an oddball 21st century phenomenon, the earliest "rogue" taxidermist, Walter Potter, was creating little anthropomorphic taxidermy dioramas with everything from rabbits to hedgehogs while Victoria was still on her throne.

But taxidermy was also an art of more general naturalists. Field work often consisted of taking a rifle or shotgun into the wild and killing animals to take back as museum specimens. Even the 26th president of the United States, Theodore Roosevelt, taxidermied his own specimens starting in childhood, and never lost his enthusiasm for it: while on a visit to a national park during his presidency, he caught, killed, and preserved a mouse he saw along the trail because he was convinced it was an undiscovered species.

The 20th century saw a cultural shift away from taxidermy as an art. Museums began replacing their dioramas with more electronics-heavy educational displays. Animal rights activists protested hunting for any reason, and fur and other animal remains were increasingly seen as dirty and germ-ridden. And as children spent more time playing indoors with video games and other distractions, they had less exposure to the realities of life and death in nature. Ultimately, taxidermy became relegated to the realm of hunters and strange family heirlooms.

Not that that stopped taxidermists. Major organizations like the National Taxidermists Association (founded in 1972) served as providers of news and issues surrounding the art, and their yearly competitions continue to bring together some of the most impressive taxidermy art in the world. They've been joined by more recent organizations like the Minnesota Association of Rogue Taxidermy, started in 2004 for the promotion of more off-the-wall styles of taxidermy that would have made Walter Potter proud.

In recent years there's been a revival in the interest of taxidermy, both for collectors and artists. Museums are breathing new life into old mounts; one recent example is the collection of striped hyena mounts at the Field Museum in Chicago, who received thorough cleanings and a new diorama after a highly successful crowdfunding campaign in 2015. Classes on making tiny taxidermy out of domesticated rats and mice abound, and books, websites and workshops demonstrate how to work with larger, more complex hides, all from the comfort of your own home. Traditional taxidermy schools are still around, some studios take on apprentices as they always have, and there are more resources for aspiring taxidermists than ever.

Renaissance Fairs, Mountain Men and Powwows

While the dominant culture in America was doing its best to make taxidermy an endangered species, taxidermists weren't alone in their stubborn dedication to dead critters. Far from the studios populated with deer heads and bobcats lounging on tree branches—and the remaining haute couture designers doing their best to sell fur coats and stoles despite the threat of red paint-wielding activists—pockets of enthusiasts wove hides and bones into their cultural interests.

Go to any Renaissance fair and you're likely to find more than one vendor selling a plethora of animal hides, fox tails and even a selection of bones now and then. Some of these end up incorporated into period costuming, while others are draped over chairs, tent walls and beds for decorative purposes. Hides were certainly a significant part of medieval and Renaissance Europe, and many fairgoers feel that faux fur just won't cut it, especially when they may have laid down thousands of dollars to create an otherwise authentic outfit. The species featured in period costuming may not always be completely authentic—there wasn't a coyote hide found in Europe until well after the Renaissance began. But the dedication to recreating the look and feel of an older time means real hides and bones abound at these events.

Not every bit of fair lore is purely historical. Hang around long enough, and you might hear someone mention a "flea fur". In Renaissance Italy it was fashionable for women to wear the hides of martens and other weasels around the neck; these were called *zibellini*, or "sables". Later historians tried to explain them away as decoys to attract fleas away from the wearer's body, but neither history nor science support this—fleas are attracted to blood, not fur. Still, the flea-fur myth persists.

Also, a note on those fox tails: while there is no historical evidence that tails were a particular fashion sensation anytime in medieval or Renaissance Europe, they've become a staple of modern fairs. Opinions of their symbolism vary, but they're generally some variation on "tail worn on the left means you're single or in

14

an open relationship, on the right means you're taken or not looking". In other words, it's Vulture Culture meets the hanky code.

While some people are in one corner reproducing a more sanitized, less plague-ridden version of medieval Europe, over yonder there's another crowd doing their best to relive the lives of the European-American mountain men who trapped furs and traversed the wilderness. Starting in the 15th century, less than 100 years after Columbus stumbled upon the New World, France began sending men to trap and hunt animals for their furs in what would one day become Canada. Other countries followed, and by the height of the fur trade in the mid-1800s millions of hides were exported every year. Since animals grow their thickest fur in cold weather, the trappers were able to pull their best profits by enduring the frigid winters while they checked their trap lines and prepared hides for transport to civilization.

Because they had such an isolated life, the mountain men would meet at a yearly rendezvous; if you didn't show up, it was assumed you were dead. This concept has rolled over into reenactment circles; people recreating the trappers' lifestyles in the 21st century are able to attend a variety of "rondys". Similar to Renaissance faires, these events are centered around period tents and other structures, with attendees in appropriate attire. Black powder rifles and other weaponry and tools used by the original mountain men are also popular.

And, of course, there are pelts everywhere; in fact, some reenactors are also actual trappers and hunters. Few live completely off-grid like modern mountain men, but a number of the hides you see at black powder events were caught and tanned by the people displaying them. They may understand a little more thoroughly than the average renaissance fair attendee just how much work goes into producing a fully tanned hide or cleaned and carved bone utensil.

While Vulture Culture's demographic does tend to be fairly heavily slanted toward Caucasians, there are a fair number of adherents who have some indigenous American heritage. Native American powwows are celebrations of cultures that have existed in the Americas for hundreds, even thousands of years.

As in other parts of the world, furs and hides have been important to Native cultures of the temperate and Arctic regions of North America. Contrary to Hollywood and other stereotypes, not all indigenous Americans make use of buckskins, feathers and beads in their clothing and other creations. In some areas of Central and South America, as well as the American southwest, wool has traditionally been an important fiber, and cultures all across the Americas have made use of bones and antlers for carved utensils, weapons and other everyday items. All of these are often paired with plant fibers, stones and other natural materials to create a glorious fusion of colors and textures.

So a visit to a public powwow can be a pretty amazing display of Native art involving animal remains. I particularly recommend watching the dancers, whose elaborate regalia results from many hours of artistic creation and generations of tradition. And there are usually vendors selling a variety of pelts and art made from animal remains.

The inclusion of powwows and Native culture in general in the discussion of what has inspired Vulture Culture may be controversial, since so many Vultures are of Caucasian descent. There's not as much discussion of cultural appropriation in Vulture Culture as in some other areas of society like academia or fashion. Certain designs that have been treated as fairly generic, like dreamcatchers, may be created by Native and non-Native artists alike, but there's debate as to whether the latter group should drop the dreamcatcher entirely. Other creations, such as fur bags and full hide headdresses, can be found in cultures worldwide, but due to stereotypes may be assumed by outsiders to be "Indian art".[2] At the same time, I would be remiss if I didn't include these cultures and their arts as having an influence on some of the things I've seen my fellow Vultures create, and contributing to the general interest in hides and bones as more than morbid curiosities.

Vulture Culture is Born!

So why talk about all these seemingly disparate groups—taxidermists and modern mountain men, indigenous Americans and Renaissance fair players? Because all of them have influenced the modern Vulture Culture, and all of them have found a social and organizational platform on the internet. In this milieu of ideas, boundaries between groups and interests became much more permeable, particularly as people thousands of miles apart share ideas. From chat rooms to forums to Facebook, today your average Vulture has access to thousands of fellow enthusiasts.

I personally consider the birth of modern Vulture Culture to be the creation of the Fur, Hide and Bone (FHB) community on Livejournal. Opened by user MintakaWolf on January 10, 2004, it eventually drew together over a thousand people who were "interested in 'dead things'."[3] People traded both ideas and specimens, showed off pictures of new acquisitions, and asked questions on everything from tanning to legalities. For several years it was the best spot to indulge your interests in animal remains.

Sure, Taxidermy.net's venerable and incredibly helpful forum came first, birthed in August 1998.[4] But it was (and still is, to a great extent) largely the domain of traditional, and predominantly male, taxidermists. FHB couldn't have been more different. Fueled by Livejournal's geek-friendly format, it attracted a younger, more diverse crowd, and the majority of active members were female (more on that in a moment.) While professional taxidermists were in attendance, most members were of a more home-grown DIY sort. And we were as likely to be interested in fan fiction, Renaissance fairs and volunteering at wildlife rehab facilities as in hunting, trapping and fishing. (Some members were even involved in all of the above!)

Over time, those Fur, Hide and Bone fans (affectionately known as "Fur,

---

2 In the twenty years I've been creating hide and bone art, I've lost track of how many times someone has come up to my booth at an event and exclaimed "Oh, look, Native American stuff!" and I've had to explain to them that no, I'm a white girl from the Midwest.

3 http://furhideandbone.livejournal.com/profile

4 http://www.taxidermy.net/forums/

Hide and Boners") could be found on every social media platform that gained momentum. Today, photo-friendly sites like Tumblr, Instagram, Twitter and deviantArt attract particularly large fandoms. Hashtags like #vultureculture and #taxidermy are chock-full of everything from snapshots of roadkill and raw, fleshy skulls to immaculately groomed fur rugs and preserved wet specimens in artistic glass jars. The ability to take a good photo seems almost as important as the hides and bones themselves.

More importantly, this new breed of hide and bone enthusiasts includes a lot of shameless nature-lovers. Many would describe themselves as environmentalists or conservationists, and made that a big part of their identities. And where traditional taxidermy often has a pleasingly wild feel all its own, Vultures are as likely to draw on the style of Victorian parlors as hunting lodges. We've drawn together tired-out stereotypes—refined femininity and rustic masculinity, urban flash and rural earthiness—and blended them into a spectrum of creative sensibility that transcends gendered expectations. We want our collections to look good, but we also want to have some connection to where they came from.

It's this emphasis on specimens that are both aesthetically pleasing and naturally evocative that defines Vulture Culture. While we can appreciate the durability of a pair of leather boots, we're probably not going to hang a pair of Doc Martens on the wall. We want professional taxidermy that looks like the animal did in life, or at least a reasonable facsimile thereof. [5] Our idea of fine art tends toward carefully crafted curiosities and feather-bedecked wall hangings. We sift through thrift store racks for vintage furs, and who doesn't love weird dried piranha from dusty old tourist traps? It's entirely too easy for our homes to end up overtaken by hordes of dead critters—but hey, they make for great photos to share with our fellow Vultures!

Lois Lane, Girl Taxidermist

One of the most remarkable things I've found about Vulture Culture is that there are more women than men involved. Traditional taxidermy has always been largely male-dominated, much like hunting, its close counterpart. While more women have gotten involved in both hunting and taxidermy in recent years, they can still both be boys' clubs, both in numbers and in culture. Girls tend to be discouraged from doing anything that's messy or gross, or which involves death (other than cooking carefully packaged meat from the store), and so most of us never have a chance to develop an interest in dead things.

From its beginning, Vulture Culture has been a haven for women and girls who refused to fall into line. We've created a space for each other in which our sex and gender don't conflict with our love of hides and bones, a refuge from those

---

5 There's a certain sort of Vulture who revels in really bad taxidermy, whether it's kitschy creations like tiny white mice having sex while wearing Victorian clothing, old and moth-eaten mounts straight out of a horror flick, or merely poor-quality Frankencritters whose amateur status is rank indeed. Kat Su's book *Crap Taxidermy* may be of interest if this is your thing.

who see our interests as "unfeminine." And we don't feel constrained by traditional taxidermy trends, which tend toward lots of snarling predator rugs and huge-antlered bucks.

Take rogue taxidermy, for example. The term was coined by artist Sarina Brewer, one of the co-founders of the Minnesota Association of Rogue Taxidermists. The official MART website, http://www.roguetaxidermy/com, defines it as "A genre of pop-surrealist art characterized by mixed media sculptures containing conventional taxidermy materials that are used in an unconventional manner." So where a traditional deer shoulder mount might have the buck looking regally off in the distance, maybe with a few fake tree branches for nature-inspired décor, a rogue taxidermist might turn the animal into a unicorn, or dress it up in steampunk costumery. While only one of the three founders of MART is female, women have stormed the genre with a vengeance. Artists like Katie Innamorato, Divya Anantharaman, Amber Maykut and Darien Baysinger are just a few of the leading ladies of rogue taxidermy.

And a group of largely female artists like Katie Clark and Ellie Willingham have subverted traditional taxidermy by eschewing hard foam forms and instead turning hides into cuddly stuffed animals. This soft-mount taxidermy still allows the animal to look more or less lifelike, but soft filling and pose-able wire skeletons make them more touchable and travel-friendly.

Keep looking through the artistic creations of Vultures, and you'll find decorated and carved bones, painted feathers, dolls and stuffed toys with real animal skull heads, jewelry made with human teeth, and a plethora of other strange and wonderful creations offered by female artists. Many of the hides, bones and other specimens available for sale as-is are also prepared and sold by women.

But it's not just about us. Birthed from subversion, Vulture Culture is equally open to anyone, whether female, male, or outside that binary. We're united in our appreciation for things that many people find distasteful, and many of us share the experience of having grown up as "that weird kid." Unsurprisingly, many Vultures are unconventional in other ways; we have a lot of goths, pagans, anime fans and LGBT folks in our ranks. While we're not some big, happy family and there are certainly conflicts among individuals, generally speaking if you like dead stuff, you're welcome to join us. (Don't know where to find fellow Vultures? Chapter 6 has more details on that topic.)

A Vulture Culture Q&A

The following are some of the more common questions I've gotten about Vulture Culture. You can ask other Vultures the same questions and likely get some varied answers, so don't take them as the final word on the matter.

Why do people join Vulture Culture?

All sorts of reasons! A lot of Vultures love nature and want to collect specimens to remind them of this wonderful world. Some work in the natural sciences and bring their work home with them. Others like outdoor recreation and collecting reminders of their wilderness adventures. Artists sometimes use hides and bones as anatomical and color references in their work. Some goth-flavored Vultures are fond of anything weird and morbid. And one particular subsection of Vultures believes the animal remains and other natural souvenirs to be sacred.

Vulture Culture makes a valuable resource for everyone above. It's a great way for people to exchange information on practices like bone cleaning and hide tanning. And there's a robust trade in feathers, bones and other specimens, both on the part of individual Vultures and established natural history shops. I also can't overstate the importance of knowing you're not alone in your interests, that you're not some sicko just because you think dead things are kind of cool, and that you're in good company including everyone from artists to scientists to farmers.

Is it safe to have dead animals in your home?

If they're processed correctly, yes. Unlike meat, you don't need to keep taxidermy and nice, clean bones in the freezer to keep them from rotting, and anything that's well-preserved can be on display out in the open. Processing and cleaning methods like those in Chapter 4 kill off the bulk of pathogenic microbes so you shouldn't have to worry about them being all germy.

A couple of specific topics: one, old taxidermy (from the Victorian era well into the 20th century) was often prepared with arsenic as an insecticide, so you want to be really careful handling it. While I haven't tried them myself, I've heard people recommend swabbing suspected mounts, especially those with white powder visible on them, and testing the swab/powder with an arsenic test from Weber or Machery-Nagel. Some suggest care when handling any mount prepared prior to 1980. You're not going to die if you have an old deer mount on your wall, but it may be best to handle older mounts with disposable rubber gloves, and as little as possible. (Wearable fur like coats and stoles aren't going to have been treated with arsenic for obvious reasons.)

Also, I've had people ask me whether you can be allergic to fur in the same way you can have allergies to live animals. Animal allergies are usually caused by the dander, or dead skin cells, of the animal. Most of the flakes will wash off in the tanning process, so they're not really that much of a concern; the same goes, of course, for saliva, which is another allergen. There are people who have allergic reactions to tanned fur, either of a particular species or in general; this may be irritation caused by the hairs themselves, or reactions to commercial tanning solutions. However, most people who are allergic to live animals should not have problems with dead, preserved one.

Do Vultures always scavenge their specimens from already-dead animals?

Some do; there are Vultures who primarily do things like collecting and processing roadkill, or looking for bones from animals that died in nature. However, others buy pelts and other remains from animals that were farmed for meat and/or hides, or that were hunted or trapped. Some specialize in searching secondhand shops for vintage taxidermy and other treasures. And a few Vultures even process the remains of animals they raise themselves for food, like chickens, rabbits or cows. The next chapter will go into more detail on where to find hides, bones and other specimens.

Do Vultures enjoy killing animals?

In all my years watching Vulture Culture's birth and growth, I've yet to find anyone who actively *enjoys* killing animals, whether for food or any other reason. Some hide and bone sellers have a strictly business attitude toward the remains in their stock. However, among many individual Vultures I've seen a reverence for the life that is taken and a healthy respect for death, whether the person in question is a hunter[6], trapper or farmer, or someone who has never killed anything bigger than a bug. For myself, I began working with hides and bones in my art because I wanted to give the remains a better "afterlife" than being a trophy on a wall, and this sentiment is pretty common across Vulture Culture in general. Many Vultures find that working with the remains of deceased animals has helped them to become more comfortable with death, and their own mortality in particular.

In fact, there have been frequent debates about the irreverent treatment of these remains. Some Vultures think rogue taxidermy, like squirrels made into beer bottle holders and the like, is too disrespectful. Conversely, I've run across people, especially newer collectors who are super gung-ho about their budding hobby, who treat their collections more like Pokémon than the remains of once-living beings (gotta catch 'em all!) There is no single consensus on what constitutes the correct way to treat one's hides and bones, but most Vultures will agree that a certain level of respect is absolutely necessary to "do it right".

What about animals being skinned alive and other animal abuse?

A few years ago a video began making the rounds online of raccoon dogs and other animals being skinned alive at a supposed Chinese fur farm. While the raw, uncut footage was never released to prove it wasn't staged, one thing is for sure: skinning a dead animal is a lot easier than skinning one that's alive and struggling. A significant proportion of Vultures have skinned and tanned their own hides, and can verify that it takes a lot of skill and patience to skin an animal without accidentally cutting or tearing holes in it, getting blood all over the fur (which can ruin it), or cutting their own hands in the process. No one in their right minds would add the challenge of a moving, biting, clawing target.[7]

---

6   There's a beautiful photoessay, "I Love Animals But I Kill Them Too: Hunting Alaskan-Style" by Christine Cunningham at http://www.bbc.co.uk/news/resources/idt-sh/I_love_animals_but_I_kill_them_too_hunting_alaskan_style. It's a wonderful narrative about turning hunting from an act of domination done in violence to an act of necessity done with compassion and appreciation for the animals being killed. There may be something there that resonates with many Vultures; I recommend giving it a read.

7   Wolfforce58205 on YouTube has uploaded a pair of videos demonstrating why it's unlikely

Yes, it is always possible that someone actually is trying to skin live animals for whatever reason; whether they're in a hurry or just horrible people. And there's no way to sugarcoat the fact that animals on a fur farm living in small cages aren't going to have nearly as good a quality of life as they would in the wild. As much as we'd like to say that everyone who cares for these captive furbearers treats them with the utmost respect and care, numbers and human nature suggest that fur farmers, like all farmers, run the gamut from ethical to abusive.

The same goes for wild animals that are hunted or trapped. Ideally a hunter will get a clean kill with the first shot, but even the best hunters sometimes miss the mark, and will have to follow an animal that bleeds to death. Most hunters abhor wasting an entire deer just to get the antlers, but there are those who give the rest a bad reputation, dumping the carcass to rot in an out of the way place. As to trapping, while newer traps are not as brutal as the old sharp-toothed bear traps of Warner Brothers cartoons fame, some Vultures are uneasy with the idea of an animal being stuck panicking for hours in a trap before the trapper makes it down the line to dispatch them.

Within Vulture Culture, there's a good bit of debate on the treatment of animals both captive and domestic. As I'll discuss in the ethics section of the next chapter, many Vultures choose to avoid hides and bones from animals they felt weren't treated well enough.

Are Vultures making it harder for endangered species to recover?

There's no formal research being done on the effects Vultures in particular have on the demand for critically endangered species' remains, but we can't deny our potential part in the matter. Any deaths in a species' population will have an effect on the gene pool as well as the health of the entire ecosystem. For example, the trade in the remains of Asian bats, like mummies and skulls, has caused a significant drop in their populations. American collectors have been specifically identified as one of the main groups consuming these remains, with thousands for sale online and in curiosity shops nationwide.[8]

While many of us are aware of the laws surrounding the trade in live and dead animals, there's still a frustrating amount of ignorance of those regulations as well, both unintentional and deliberate. Some species that are 100% legal to collect may not be sustainably harvested at their sources, like the aforementioned bats, and we simply don't have enough information on their populations' health. The International Union for Conservation of Nature (http://www.iucn.org), a leading

---

that anyone is deliberately skinning animals alive. They are at https://www.youtube.com/watch?v=j4rhGiyMEDM and https://www.youtube.com/watch?v=8xJfl-gSr8o . Also, the International Fur Federation claims to have interviewed the people in the infamous video, who say they were paid by animal rights activists to stage the live skinning; the interview is at https://www.youtube.com/watch?v=z6joIOEk6JU .

8   See http://www.newsweek.com/thousands-bats-slaughtered-annually-asia-end-ebay-and-etsy-artsy-americans-681147

authority on species endangerment and extinction, has a Red List of species that they consider threatened or endangered, but there are thousands of species that they haven't yet assessed, so we can't know for sure if our collecting of their remains is truly impact-free.

That being said, Vulture Culture is pretty good at self-policing. If someone on a forum asks to buy something that's not legal, or at least questionable, there will almost certainly be someone in the very first few comments saying "Hey, you actually can't have that, and here's why". Unfortunately there will always be ignorant and obstinate people whose attitude is "Well, screw you, I'm going to buy/sell/hunt/have this anyway." And it's my hope that the next chapter of this book, along with my animal parts laws database online, will help increase understanding of the legalities and the reasons behind them to reduce that sort of destructive thinking. That way as a community we can be more responsible for the ecosystems we draw our specimens from.

Regarding the definition of "endangered species", I'd like to add that just because a species is considered endangered in some places doesn't mean that all of its populations are threatened on a widespread scale. Gray wolves are a great example—environmental organizations (including many that I support) often bring up the gray wolf as an endangered species needing protection. In most places in the world, I agree; in the lower 48 states, for example, wolves have only recently begun to return to places where they were exterminated up to a century or more ago, and hunting is prohibited in most of those states. In Canada and Alaska the wolf populations are steady enough that limited amounts of legal hunting are allowed, and it's legal to trade in their remains. (Whether you're willing to buy a skull from a hunted wolf or not is something I'll get into in the ethics section of the next chapter.)

Do Vultures only collect animal parts?

While we're best known for our collections of feathers, fangs, bones and fur, Vultures often also collect other bits and pieces of nature. My own collection includes fossils and other mineral specimens, dried plants and fungi, and art made from all of the above. I also love old natural history books, and posters of scientific illustrations. Some people even include found objects like old railroad spikes, antique glass bottles and 19th-century photographs to be relevant to Vulture Culture. Each collection is going to be different, so don't feel you have to have a particular thing to be part of the in crowd.

Do you have to tan your own hides and clean your own bones to be a real Vulture?

Nope! These are certainly good skills to have, and there are plenty of resources for people who'd like to process raw materials into more finished specimens. It's not a bad idea to at least understand how these things are done. But all you really need to be part of Vulture Culture is an appreciation for dead stuff. Personally, I prefer getting already-prepared specimens; I would rather be able to jump right into making my hide and bone art and leave tanning and bone cleaning to people who enjoy them.

Is there such thing as a vegetarian Vulture?

Yes, and vegan ones, too, at least when it comes to their diet. While some vegetarians and vegans abstain from animal products in general, others only restrict what they eat due to health or personal reasons. Some may be pickier about sourcing, only choosing to collect secondhand animal parts, or the remains of creatures that died in the wild, or had accidental deaths like being hit by cars or trapped in the walls of a house. But that's not absolute; I know of vegetarian and vegan Vultures who won't touch meat but will happily support an individual trapper by buying their hides or bones directly.

Can I be a Vulture, too?

Yes! As I said earlier, everyone's welcome here. You don't need a big collection of taxidermy, and you don't have to clean your own animal skulls. Even if you're broke and living in the middle of a city, a collection of cleaned chicken bones and a few pigeon feathers is more than enough to get you started. Some Vultures don't even have a collection of their own, simply enjoying the "fandom" of dead things. And it doesn't cost more than an internet connection to appreciate others' treasures. We love the attention; leave a kind compliment and you'll make a Vulture's day.

It's okay if you envy others' collections and want to start your own, however humble. Let's talk about where you can get teeth, claws and scales of your very own!

Chapter 2:
The Successful Scavenger:
Where Do I Get My Goods?

Probably the number one question I get about Vulture Culture is "where can I get bones/hides/feathers/etc.?" You have a lot more options than you might expect— but there are also some important restrictions in place, too. Before you go on that internet shopping spree or run out into the woods with a garbage bag and rubber gloves, read this chapter first.

On Buying "New" Remains

This is the most common source for specimens; while much is made of preparing the remains yourself, many Vultures don't have the space or time to macerate bones for months or skin roadkilled possums before they rot away. So there's no shame in buying things others have preserved.

The following is just a partial list of what you may find on the market:

- Hides (usually tanned, but dried/salted/"green" hides are also available for those wanting to try tanning)
- Bones (some may only be "nature-cleaned", meaning they sat outside for a long time and may still have dried stuff stuck to them)
- Feathers
- Antlers and horns (mounted or loose)[9]
- Wet specimens (partial or whole animals preserved in alcohol, formalin or another preservative)
- Mummified or dried specimens (including dried insects, among others)
- Taxidermy
- Articulated skeletons
- Study skins (birds that have been prepared for museum study, usually with the wings tightly locked against a stretched-out body)

There's also no end to the creativity produced by artists who work with the above specimens and more. Some suppliers even sell lots of damaged or second-quality items just for crafters to use.

So you have a basic idea of what's on the market—but where can you actually go shopping? Starting in your area, check for local taxidermy shops; many of them sell hide scraps, bones and other parts they aren't using. Some larger urban

---

9 There's more information in Chapter 4 about the differences between antlers and horns.

areas have specialty oddity and natural history shops carrying animal, plant, fungus, mineral and other specimens, as well as relevant books and other resources. You might even find oddball establishments like Portland, OR's Zymoglyphic Museum, which either may have their own little gift shops or can at least direct you to local oddity vendors.

Online you have even more options. Individual wholesalers and retailers may list thousands of bones, horns and leather on their websites for instant purchase.[10] Some of them also maintain virtual shops on Etsy, eBay and similar sites. Social media sites like Facebook and deviantArt have dozens of groups dedicated to collecting and trading animal remains; some are public, though others require you to request membership; I've listed a few of my favorites in Chapter 5. Other social media sites like Tumblr and Instagram don't have groups, per se, but you can search tags like #taxidermy and #vultureculture, which some Vultures use to draw attention to items they have for sale.[11]

Your best bet is to go where the Vultures are flocking and ask them for their current favorite places to shop, both brick and mortar, and online. For example, go on Tumblr and make a post asking for good sources of whatever you're looking for, and then add those special tags like #vultureculture. With a little time and luck you should get a nice bunch of responses, perhaps even from artists and shop owners themselves!

Secondhand Sources

I am a huge, huge fan of thrift stores and antique shops; I also have trouble resisting sales of the yard, garage and estate variety. Most of my vacations involve a significant amount of time spent poking around in shelves and cabinets full of other people's old stuff in the hopes of scoring some unusual treasures. At least I'm in good company. Many Vultures consider it a sort of victory when they manage to find an old taxidermy mount or mink stole in the dusty recesses of a secondhand shop. This event is often immortalized with pictures and bragging online, much to the delight (and envy) of other collectors.

Not every thrift shopping trip will be fruitful. There will be days when you come home with absolutely nothing, not even a tragically unstylish leather jacket. Don't fret—it can take a few visits around your collection of local shops to determine which ones are most likely to yield something nifty. Bigger businesses

---

10 You may be wondering why I don't have a massive list of professional Vultures who offer services like bone cleaning or who sell items. Since businesses enter into and drop out of the market at unpredictable rates, the list would be both obsolete over time, and would necessarily be incomplete as there are a lot more excellent businesses and artists out here than I have the space to include.

11 Again, as the internet is constantly shifting and changing, the more time passes from when this book was published, the more links will be out of date. It's likely that even a couple of years after its publication there may be new hubs of commerce for Vultures, and the ones I mention here are out of favor or even defunct. A decade ago Livejournal was the place to be, and now who's even on there anymore?

have more selection, but don't discount the smaller ones, especially if they're a bit more out of the way. They might not be as picked-over, and you could find a real gem someone else overlooked.

Even if you're stuck at home with no transportation, you can still find secondhand hides and bones online. While there are individuals who wheel and deal in new animal parts, sometimes Vultures put their personal collections up for sale, in part or in whole. This is often so they can get money to buy other specimens they've had their eye on, or to get a bit of cash to cover this month's bills. Other times it's a simple matter of de-cluttering. So I generally don't feel bad about buying something from someone's private stock, unless it's something very dear to them that they're sacrificing so the rent gets paid.

Finders, Keepers

If you don't have a ton of money but plenty of time and energy, you may prefer to go scavenging for bones and other remains in the great outdoors. The easiest option is to look for fallen feathers and weathered bones that are already more or less clean and there for the taking. Woods and fields are great places to look, though you'll want to get permission to scavenge on private property. When I lived in rural areas I would drive around the backroads for fun, and occasionally would find places where hunters dumped carcasses or unscrupulous pet owners left behind the remains of dead dogs and cats that they didn't care to bury or cremate. Sad though these were, they were an opportunity for me to collect bones for art and personal use.

There's a lot of open space out there. How do you know where to actually look for remains? Sometimes it's just sheer luck, but there are certain places that are especially likely:

- Rivers, streams and creeks all may harbor bones of animals washed down in flooding or frozen in winter. Look in the outer edges of bends or on embankments, especially in slower-moving parts of the waterway. Downslope areas in general are also good places to check, since gravity may cause carcasses to roll or slide downhill, and animals may find it easier to drag them to cover there so they can scavenge in peace.

- Look for pastures where cattle and other livestock are kept, especially in winter when they may succumb to cold, starvation or predation.

- If you can observe a place over a matter of weeks, see where the wildlife like to congregate and travel. They're more likely to die there, and predators are more likely to leave the remains of their kills. Just don't intrude so often that you disturb them; avoid nests and dens, especially in spring and summer.

- A lot of it is just a matter of practice and developing a good eye. Look for any unusual patches of white or light tan in the grass and bushes, or shapes that look like bones. A lot of the time it'll be a leaf or paper or something that is definitely not a bone, but the more you look the more you'll be able to discern whether something is worth investigating.

Because dead animals begin to rot almost immediately, it's very hard to get a hide from an animal that died of natural causes. You're better off hoping for some bones to clean.

With regards to insects and other invertebrates, you're of course going to have to look more closely for your specimens. In this case you may actually be better off looking indoors, especially if you want bugs that are already dead. Windowsills, corners and behind furniture are all good places to look. If you want to search around outside, you can turn over rocks and logs, but please be sure to put them back exactly where you found them as they are crucial wildlife habitat. Spider's webs can occasionally yield the wings of butterflies and moths, though be sure the spider has finished dinner first. You can also search the banks of streams and rivers for invertebrates that have washed up naturally, or that have died after mating. Look for the shed exoskeletons of crabs, crawdads and other crusteaceans, too. Many invertebrates die off in fall, so you may have better luck hunting for their remains then.

If seashells are your thing (whether they're from oceanic or freshwater gastropods), collecting them is simply a matter of keeping your eyes peeled along the banks of waterways. You may also stumble across the shells of more landlubbing molluscs like snails walking across your lawn or through the woods on a cool summer evening.

While you're out there stomping around the woods and fields, please practice good situational awareness. Pay attention to where you're walking so you don't end up falling off a cliff or running into a patch of poison ivy. Keep an eye on the weather, especially if extreme weather like thunderstorms or heavy snow is possible. If you're going to leave established trails be very sure you can find your way back to your starting point; take a compass and a map if you're headed into wilderness areas where your phone may not get enough signal to bring up directions. If it's hunting season wear an orange vest or other obvious and bright clothing; on some public lands it may be the law for you to wear blaze orange at that time of year. And it's a good idea to tell someone where you're going, especially if you're headed out alone.

Be aware that some states have regulations about picking up the remains of particular vertebrate species, including protected ones and game animals, as well as some seashells. Many federal, state and local parks prohibit the collection of *any* natural materials even for personal use. Why is that? After all, don't the parks belong to the people? Well, in theory, yes. But park services are in place to make sure we don't love our parks to death. We aren't the only ones who want shed antlers and dry bones. Rodents and other creatures chew on these to get much-needed calcium, and as they decay further they feed the soil, and the animals, plants, fungi and microbes that get their food from it. The same goes for fur, feathers and flesh. The more we take out of an ecosystem, the harder it has to work to replace the missing nutrients. It may be best to just take some of the bones you find, and leave the rest for nature.

That's in a wild ecosystem, though. What about roadkill? After all, it's likely to just rot by the side of the road, or get thrown into a landfill by the local sanitation workers. On the surface this seems like a great resource to use. But it's messy, smelly, and can be dangerous to collect, especially along busy roads. There may also be laws prohibiting roadkill pickup in your area, though some states have a salvage permit you can apply for.[12] On a related note, railroad tracks are owned by the companies that build and maintain them, and it is possible to get arrested and charged with trespassing even if you were just picking up bones along the tracks.

If you are going to collect roadkill, be prepared to do a lot of work to preserve the remains. Chapter 4 will go into more detail about some of the processes involved. Many Vultures find the effort worth it, both for the feeling of personal accomplishment, and knowing the remains were treated respectfully along the way.

Should you happen to find the carcass of a cat or dog struck by a car, be aware that it may be someone's pet and they're looking for it. Even if it's thin and dirty and doesn't have a collar, it may have simply been lost for a few weeks. You can take it to a veterinarian or animal shelter to have it scanned for a microchip, or buy a scanner yourself if you pick up a lot of roadkill. Keep an eye on lost pet ads online and in local papers, both from the past few weeks, and for the next few weeks. If you have a dedicated roadkill freezer, store the animal there until you're really sure no one is going to claim it. Some people prefer to bury dogs and cats; if you do decide to salvage the remains be aware that it is illegal to buy or sell dog or cat hides, though bones are just fine.

Also, make sure you collect roadkill safely. Stay off the road itself and be aware of traffic headed your way. In fact some of the best finds may be a little ways of the road, as animals struck by cars may run off into the woods before dying, though they may not make it very far. One way you can prevent further roadkill is to drag carcasses off the road and into surrounding vegetation. That way scavenging animals can safely enjoy their meal, and the leftovers become food for the plants.

Wear rubber gloves and put everything you get in heavy plastic garbage bags; you may also wish to wear a respirator mask while collecting, and if you have to have your "treasures" in the cabin of the vehicle with you. Wash your hands VERY well afterward; you may consider just taking a full shower to be extra safe.

Why do you need to worry so much about cleanliness, especially with fresh carcasses? Well, there are some zoonotic diseases that can be passed to humans from other animals, and you generally can't tell whether a carcass is infected just by looking at it. Rabies is the most infamous one found in many mammals; it's spread through saliva and fatal if contracted and you don't get a series of shots of post-exposure prophylaxis shortly thereafter. However, there are a number of viruses, bacteria and fungi that can survive on a carcass for varying times after death, like

---

12 You can find out more about laws in your area at http://www.thegreenwolf.com/animal-parts-laws - not every area may have roadkill laws listed, so your best bet is to contact your state fish and wildlife department.

Cryptococcus found on pigeons and other birds, or leprosy in armadillos. If the animal is very freshly dead you'll need to be mindful of ticks, fleas, mites and other external parasites. And there are plenty of pathogens that will give you gastrointestinal distress if you don't wash your hands well and then go eat something.

The most dangerous time to work with fresh carcasses is when they still have liquid body fluids, because the animal has only been dead a short time, and the fluids are messier. When you first bring your prize home, hose it off outside or in a utility sink if possible. Bleed it out if need be, letting the blood drain safely away from the home (or with a smaller animal directly into the sink drain.) If you're skinning the animal, flesh and wash the skin thoroughly. Then put it into salt which will both help dry it up a bit and kill a lot of the pathogens on it. Or immediately begin the tanning process, which will also help to get rid of bacteria and the like. You can immediately begin cleaning bones, whether that's through beetles, burial or maceration; be aware that marrow can survive for months after and may still harbor a few germs, so be careful if drilling into or cutting fresher bones.

Feathers can be washed with gentle soap and water, dried, and then put in a plastic bag in the freezer for a couple of weeks to kill off any mites and other parasites. Some people also swear by soaking them for an hour in 50-50 isopropyl alcohol/hydrogen peroxide mix but I haven't tested it myself. If you get them wet you need to dry them quickly so that they don't mold. Finally, if you're preserving partial or whole wings and tails in Borax or salt, that should also kill off a lot of germs through dessication.

For shells, first make sure that the shell is no longer inhabited by a living creature. Once you get it home, let it sit in water for a week, changing the water out every couple of days. Don't bleach it, unless you want to lose all the color. Instead, lightly buff it and remove any caked-on gunk with sandpaper, and then you can use mineral oil to bring out the color more vividly.

Other Sources

If you want a super-simple source of animal bones, albeit of a limited selection of species, look no further than the local grocery store. Most will be stocked with bone-in beef steaks, chicken breasts and pork ribs. Some even carry packages of bones for making soup. If you have a butcher nearby you may be able to get an even better selection, since some cuts of meat are de-boned before going on the market. Should you be in the vicinity of a small family-owned farm that slaughters its own livestock, you could even score a few animal heads for skulls.

All of these will require some processing. However, be very aware that cooking bones significantly weakens them and can affect their color and structure. So if you're looking to create display specimens or use the bones for art, clean the meat off of them and cook it separately, cleaning the bones with beetles, maceration or burial (more on that in Chapter 4.) You may be able to salvage denser bones from cooked meat without them being too damaged, like femurs, but the same disclaimers apply regarding strength and other qualities.

With the increase in popularity of snakes as pets, more pet stores and individual suppliers offer whole rats, mice, and even rabbits and chickens frozen for snake food. Some may feel snake food is the only justifiable use of the remains. Many Vultures, though, see them as a good opportunity to practice preservation techniques without needing much space or money. You might also ask your local pet stores, especially independently owned ones, if you can collect their deceased animals. Be honest about what you're going to do with them, and offer to pay for them.

A Question of Quality

When you're shopping for hides, bones and the like, you may run across certain terms describing what condition the items are in. While there is no official set of definitions, you can use mine as a general guideline, with the caveat that different people may disagree with the details.

Museum-quality: This describes specimens that are the best of the best. They're profesionally prepared and are in excellent condition with no visible damage or flaws. Museum-quality taxidermy will look strikingly like the live animal was frozen in place, while museum-quality bones will be clean, white, and with no missing pieces or cracks, and no residual grease. A pathological specimen, such as a skull showing severe dental infection in the jaw, still counts as museum-quality even with the abnormality in the bone as it as caused by nature rather than postmortem processing.

Taxidermy-quality: This is a tanned hide of sufficient quality to create a taxidermy mount. The tan should be newer and good quality, with hide that has a bit of stretch when wet so as to not tear when being stretched over a form. The cartilage in the ears is generally separated from the skin on the backs of the ears or even removed and the ears are "turned" inside out to make tanning more effective; the eyelids, nostrils and lips are intact, and the lips are "split", which means that they are cut open from the inside edge and spread out while the flesh inside is removed so that the lips can then be wrapped around the mouth of the form later on. Holes in the hide should be minimal, and the hides are often dorsal-cut (down the back), unless they are intended for a rug, in which case they are ventral-cut (down the neck, chest and belly.)

Display-quality: These are specimens that are in very good condition, but which may have a few minor flaws. A pelt may have a few small bare spots, while a skull might have a missing tooth or two. Still, they are otherwise processed well, and many Vultures would be happy to have them on their shelves for all to see.

Craft-quality: Artists use specimens of all qualities for their work. However, craft-quality is a term usually reserves for specimens with some damage, like a hide with large holes, a feather with missing barbs, or a cracked or broken bone. The preservation may not be the best; a poorly tanned pelt may be overly greasy, or full of holes, or tear easily. They are also not always as clean as other specimens; a skull may still have some yellow grease spots and smell, for example. Some of them require extra care before being incorporated in artwork, such as gluing a fabric or leather backing to a particularly delicate piece of fur before sewing it. Some people divide craft-quality specimens into subcategories, most commonly A, B, and C. A may describe a specimen that has a small amount of damage or missing parts, B one that is more damaged but still somewhat intact, while only a small portion of a C quality specimen may be present and/or usable. Please note that some people are happy displaying "craft quality" specimens, so these terms should not be seen as having rigid definitions.

It is a good idea to ask a seller for details about something they're selling before making your purchase, especially if they haven't been especially elaborate in their description. If they say that a vintage pelt is "damaged", does that mean that it's moth-eaten and hair is falling out, or whether the hide has dried out so much as to be dry-rotted, and therefore the skin will easily tear?

It's the Law—and It's Important!

You also want to make sure that its actually legal for them to be selling what they have, and for you to buy it. I want to make something very, very clear: just because someone has something for sale, it doesn't mean that it's legal. For example, in Oregon it's not legal to sell deer antlers on an intact skull or skull plate, yet antique stores all over the state have old antler mounts hanging up for sale. Many Americans also import and export wildlife remains through the mail without obtaining a federal import/export permit and observing other U.S. Customs restrictions. Most of the time it's due to ignorance of the law, but they—and you— can still get in trouble for the sale.

There are also people who will misrepresent what species a bone or hide came from. Sometimes it's by accident, especially if it's an obscure animal, or from a species that looks very much like another. Other people will deliberately lie about the identity of what they're selling because they want to make money without getting caught. If you aren't 100% sure that what you're buying is legal, leave it be!

Some cultural artifacts are made with animal parts, and they have a whole different layer of considerations attached to them. Many of these have been stolen from archaeological sites, or directly from indigenous people. Even if the theft happened centuries ago, it still presents ethical problems. There are also laws in the U.S. and elsewhere levying pretty significant penalties for the illegal trade in artifacts, whether or not they're made with animal parts.

Further information on artifacts laws can be found with a few quick Google searches or a call to a museum or university archaeology or anthropology department. Meanwhile, I'm going to dedicate the next few pages to animal parts-related laws.

Allow me to preface this discussion by making it abundantly clear that I am in no way a legal professional, I have no legal training, and nothing in this book should be considered legal advice. I am merely a layperson who, confused by the plethora of regulations surrounding animal parts, sat down one day a few years back and created an online collection of links to the various laws to make researching them easier, for me and anyone else who happened by.

The laws I am including in this section are primarily, though not exclusively, pertinent to Vultures in the United States. Whether you live in the U.S. yourself, or trade with people who live here, you'll want to be familiar with the legal restrictions, both so no one gets in trouble, and so any international packages don't get seized and lost forever.

There's a common misconception that if the remains of a particular animal were collected before the law that now prohibits them was put into place they're still legal to trade, and people sometimes try to sell "pre-ban" everything from polar bear claws to leopard skulls. Be aware that not every law has an exception for these older items, and those that do generally require them to have official documentation proving their age. I own a juvenile sea lion skull that has an official letter and tag from NOAA verifying it as pre-dating the Marine Mammal Protection Act of 1972, and it was sold legally to me by Skulls Unlimited. However, if someone tried to sell me a sea lion skull a deceased relative found on the beach in 1954 and kept on their shelf for decades, it wouldn't be a legal sale, even with the skull's age.

I know it can all seem confusing and overwhelming, but do your best with the information you have. When in doubt, contact your state or federal game agency, or a lawyer proficient in these laws. The laws I'm about to discuss are a great starting point, both because of their broad reach and because they demonstrate how different laws can have greatly varied limitations, requirements and exceptions.

Convention on International Trade in Endangered Species (CITES) – 1975

This is the one big international set of regulations all readers need to know about. All but a few countries have agreed to follow its mandates, and essentially all primarily English-speaking countries have signed on, so chances are if you're reading this it applies to you. CITES governs the international trade in both live and dead animals and plants considered to be threatened or endangered, and protected animals may be covered under one of three appendices:

- Appendix I includes species that are completely illegal to trade in. Generally these are highly threatened animals and plants; leopards, cheetahs, chimpanzees, rhinoceroses and pandas are just a few of the better-known examples.

- Appendix II covers species that may or may not be endangered, but which would be seriously affected by overhunting and other population loss. Some legal trade is allowed, but international trade requires official CITES paperwork declaring the species and source. Gray wolves, most African lions, African gray parrots and black bears are a few of the many Appendix II species.

- Appendix III is reserved for when a country wants protection specifically for its own population of a species and asks other CITES participants to help restrict or prohibit trade in animals or plants from that population in particular. The two-toed sloths of Costa Rica and alligator snapping turtles in the U.S. are two such populations of otherwise unprotected species.

Lacey Act - 1900

This was the first major law to regulate trade in wildlife and plants in the U.S. It provides penalties for interstate trade in wildlife or plants that violates any federal, state or Native American tribal law. It's not as well-known as some of the other laws, but it is still frequently used to prosecute poachers and other purveyors of black market live and dead animals and plants.

Migratory Bird Treaty Act – 1918

Did you know that if you pick up the feather of almost any wild bird in the U.S. you're violating federal law? How about if you remove a robin's nest next to your bedroom window because the birds wake you up too early in the morning? The Migratory Bird Treaty Act prohibits the possession or trade in any species of native bird that migrates between the U.S. and Canada, Mexico, Japan, and/or Russia. This includes the live birds, any of their remains (even naturally molted feathers), their eggs and nests, and makes illegal the harassment of the birds themselves.

The law is strict for good reason—in the late 19th and early 20th centuries, many bird species had been decimated by too much hunting, and by the demand for feathers (and whole bird skins!) for women's hats. That's when we lost several species, most famously the passenger pigeon, which in the space of just a few decades went from numbering in the millions to complete extinction.

There are exceptions; domestic birds like chickens are fine to possess, as are invasives like European starlings, ring-necked pheasants and house sparrows, and terrestrial game birds like turkeys. Waterfowl such as geese and ducks are a trickier prospect; even if you hunt them legally, you're pretty limited as to how you can pass on the remains to anyone else.

Bald and Golden Eagle Protection Act – 1940

This is the famous "eagle feather law" that levies large fines of $100,000 or more for possessing even a single molted feather from a bald or golden eagle, along with other remains, live birds, eggs and nests. It also prohibits any harassment of eagles, or any activity that may disturb nesting sites. Exceptions may be made for enrolled members of federally recognized Native American tribes, or scientists needing research specimens, as well we wildlife rehabilitators. The only legal source of eagle remains is the National Eagle Repository near Denver, Colorado, and it can take years for a formal request to be fulfilled.

Endangered Species Act – 1973

This is probably the best-known of the U.S.'s animal-related laws. Like other laws it prohibits the live capture, killing and trade of protected species of wildlife and plants. However, it is also instrumental in protecting the places where these animals live, as the law can prevent construction, mining and other destructive practices in critical habitat. Unlike most other federal laws, the Endangered Species Act includes both local populations and members of the same species permanently living outside of the country.

Marine Mammal Protection Act – 1972

As the name suggests, the Marine Mammal Protection Act governs animals ranging from seals and sea lions to whales and dolphins, along with polar bears, manatees and dugongs, and sea otters. It makes it illegal to possess these animals live or dead (to include all remains) and also to harass or otherwise threaten them. Alaskan Natives are allowed to harvest and create art and other items from some protected species like the sea otter, and they can sell completed crafts to any U.S. buyer, though said buyer cannot then go and resell the items. Scientists may also apply for research permits. NOAA has been known to issue permits for verifiable pre-ban specimens, such as old scientific specimens.

Dog and Cat Protection Act - 2000

For years dog and cat fur, particularly out of China, has been mislabeled as wolf, rabbit and other wild pelts. This law makes it illegal to sell domestic dog and cat fur within the United States (several states have their own prohibitions as well). This does mean you can't sell the hide from the feral cat that got hit by a car outside your home, but you can turn a customer's dog into a taxidermy mount since all they're paying for is the labor and taxidermy materials, not the hide itself. It also doesn't affect other parts of the animals, such as bones.

U.S. Customs Restrictions

This isn't a single law, but a whole suite of them. The short version is that shipping wildlife parts outside of the U.S. or importing them here from elsewhere isn't as simple as slapping a customs form on the box and calling it a day. You need a federal import/export permit, and there are inspection fees on top of that, so legal shipments can be an expensive prospect. Not following these regulations can result in everything from your package never arriving at its destination, to federal charges.

If you want to look up more details on these and other laws, my online database is located at http://www.thegreenwolf.com/animal-parts-laws. I've compiled the laws above, along with laws from each of the 50 states, and as many non-U.S. laws as I was able to compile. Again, this resource is offered for your research, and is not legal advice.[13]

Legalities Checklist

All those laws may seem pretty overwhelming! I've created this checklist to help you determine whether something is legal for you to buy, sell or possess. When in doubt, or if you want to seek a permit to have something that is otherwise banned, contact your local fish and game authority. And remember that even if a specimen pre-dates a particular law, you need to have solid paperwork proving that fact, because the burden of proof is on you.

- Is this specimen banned or restricted by the Convention on International Trade in Endangered Species? Remember that almost every country in the world abides by CITES.
- Is this specimen banned or restricted by federal laws in your country? You can find federal laws from several countries, not just the U.S., at the link above. Remember that some laws may cover many types of animals, such as the Endangered Species Act or Lacey Act, while others are more specific, like the Migratory Bird Treaty Act or Marine Mammal Protection Act.
- Does your state or province have a ban or restriction on this specimen? Even if CITES and federal laws don't affect this specimen, your state law will still be in effect.
- Are there any local laws in effect? For example, as of this writing California cities San Francisco, Berkeley, and West Hollywood have laws banning fur sales.
- Even if the specimen is legal to buy, sell and possess where you are, if you intend to ship it elsewhere is it legal for the person you are shipping it to to have it? For example, some states in the U.S. have restrictions on certain parts of deer due to the risk of spreading Chronic Wasting Disease.
- Even if the specimen is legal for you and the person you're shipping it to to have and trade, do you require any special permits to ship it internationally? For example, U.S. Customs has different regulations depending on what type of animal part it is (a tanned hide versus body fluids), or whether it's a domestic animal versus wildlife.

---

13 Should you happen to know of a law that is not currently on the site, or have further information concerning one that is, please contact me at lupa.greenwolf@gmail.com with a link to the law or information. Hearsay isn't solid enough evidence; I need an official website or well-researched article.

- If a new law is passed concerning certain species, does that law affect any of your collection with regards to possession, trade, or moving it across state or country borders?

If after asking all these questions it looks as though the specimen in question is a-okay, you should be alright. You are still encouraged to talk to other Vultures about their experiences, and as always if you have any doubts legally contact your local game authority.

One more note on legalities: the Vulture Culture tends to be fairly self-policing, at least online. Generally speaking if someone posts something questionable or downright illegal in a space with other Vultures, they're going to be an earful about it, as well as advice on how best to dispose of it. Some of us also like to inform antique shops and other entities out in the world when they have something they shouldn't, as they often are ignorant of the sometimes obscure laws.

My personal tactic as someone who's not in law enforcement is to educate rather than report, unless it's something obviously deliberate like poaching. Most of the time, though, what you get is someone who was given a gift made of something illegal, or who picked up a migratory bird feather off the ground, and who has no idea they're breaking the law. They're almost always going to be grateful that you told them and get rid of whatever they have. Calling wildlife authorities in these cases isn't just a waste of the authorities' time and attention when they're already stretched pretty thinly, but it's also overkill. I like to give people the benefit of a doubt, and also not ruin their life for making an innocent mistake that even a lot of newbie Vultures have made.

Detangling Ethical Conundrums

Legalities aren't the only reason you want to be sure of your specimens' sources. Whether you're open to hide scraps from farmed foxes, or will only collect bones from animals that died of natural causes, it's important for you to determine where your personal ethical boundaries are. Every Vulture decides this for themselves, and often our boundaries shift over time as we gain new information or reassess our relationships to the animals whose remains we handle. There's no single right answer; the only thing pretty much every Vulture will agree on is "Don't deliberately cause pain and torment to animals just because you want them to suffer". Beyond that, here are some possible stances to consider:

Equal-Opportunity Collector: You aren't picky where your hides and bones come from—farmed or hunted, roadkilled or antique, you'll happily take them all on. You may feel that they're dead anyway so it doesn't especially matter. Or you could believe that all animal remains deserve respectful treatment no matter the manner of their death. Farmed fur may not be an issue because you want to financially support the small farms that raise the foxes, mink and other animals, though you want the farms to follow regulations to make sure the critters are well cared for and killed as humanely as possible.

Wild Ones Only: You aren't a fan of commercial fur farms since you don't think wild animals should be kept in captivity, at least not for the purpose of harvesting hides. However, you're fine with pelts and skulls from animals that lived out in the great outdoors. You might make exceptions for secondhand items like vintage mink stoles that were likely made from farmed animals several decades ago, but you're probably more comfortable with roadkill.

Accidents Happen: Speaking of roadkill, some folks will only collect parts from animals that died accidentally. Since these carcasses often get thrown into landfills, you'd rather not see them go to waste. You'll also pick up other animals that died accidentally, like pigeons that flew into windows, but might skip that rat that was poisoned on purpose. Nature-found remains are likely okay with you, unless they're deer carcasses dumped by trophy hunters who just wanted the antlers.

Eat It and Wear It: If this is you, then you're okay with the remains of any animal that was eaten by humans. That could be anything from wild deer and elk to domestic rabbits and chickens. People do eat bear and alligator meat, too, so you have some variety to work with. You may even hunt your own deer and make use of the entire carcass. Some people in this camp won't eat conventionally farmed animals, only free-range, as part of their ethical choices. A few only use remains from animals they personally killed and prepared.

Nothing New Under the Sun: You're strictly secondhand. If it came from a thrift store, flea market, or some farmer's barn wall, you're all over it. You may not be okay with tails and other scraps left over from garment manufacture since they're technically new. Some really strict adherents only buy items that are decades old, like vintage furs and taxidermy.

Nature-Found Only: You want nothing to do with the commercial fur and bone trade, even secondhand. The only remains you'll allow yourself are those of animals that died naturally, either as prey for other creatures, or through disease, injury, old age, etc. You might flex a bit when it comes to old hunting dumping grounds or roadkill, depending on the situation.

Protect Endangered Species: Sure, there are animals that are legal to hunt and sell, like African lions and most species of fruit bat. But you feel that if they're considered endangered, especially if they're on CITES [Convention for International Trade of Endangered Species] or the Endangered Species List, you shouldn't contribute to the stress on their populations. You may make distinctions for different populations: you won't buy a wolf skull from Idaho, but Alaska's wolf populations are healthy enough that you're not worried.

Everything's Fine—Except Us: Yes, it is legal to trade in human bones and other specimens in many places, the United States included. There are surprisingly few laws covering anything besides outright theft of human remains, and so it's relatively easy to get skulls, ribs, femurs, and whatever else you like. The ethics are another matter entirely; few human remains have a paper trail explaining exactly where they've come from and been since then. Rumors abound of political prisoners being killed so their organs can be harvested for the black market, and the demand for bones has created even more of a potential profit. Even vintage displays from medical schools are suspect, since many of these were stolen cadavers supplied by gravediggers, and others may have been taken from indigenous burial sites. So you may find yourself more than willing to possess the remains of any legal animal—except *Homo sapiens sapiens*.

I Don't Give a S%@*: This is the only ethical stance that I will condemn. There's a small portion of people who feel that legal and ethical considerations don't apply to them. They may become defensive any time someone points out why their collection is a problem. Or they might feel that since no one has caught them yet, they can keep acting with impunity. These are the people who have no qualms about driving a species to extinction, just as long as they get their trophy. It's an intensely selfish and short-sighted view to take, and I actively encourage you to avoid it at all costs.

With the exception of the last one, don't worry if none of these options is a perfect fit—they're just examples to help you think about your ethical boundaries. Take note of what details apply to you and which ones you don't agree with and why. As you read through this book, and then continue your adventures in Vulture Culture, when you run across a new situation or question keep asking yourself "Is this something I am personally okay with?"

One more thing: although ethics are largely personal matters, there will be discussions on what's considered "good" ethics. These discussions can get really heated sometimes. If you choose to wade into such a debate, remember to keep your cool, talk primarily about your own stance rather than attacking others', and politely but firmly back out when you feel the need. These debates expose everyone to new and different ideas, but if they turn into massive arguments they can become counterproductive very quickly. (For more tips on diplomatic relations in and around Vulture Culture, check out Chapter 6, "Vulture and Our Neighbors".)

The Cruelty-Free Conundrum (and Other Labels)

Head to Etsy, eBay, or any other online seller of dead things, and you'll frequently see terms like "cruelty free", "humanely sourced", and so forth. They're kind of like the word "natural" in the food world: there's no single clear meaning. And if you ask ten sellers what their definition is, you're likely to get ten different answers.

So why use the terms if they don't have commonly agreed-upon definitions? Well, a big part of it is because Vulture Culture has gotten so much backlash from animal rights activists and other critics. I can speak from experience that it can be incredibly stressful and even scary when you get a bunch of people yelling at you either in person or online, telling you that you're an evil murderer and that someone should kill you and skin you.

One of the ways to keep such characters at bay is by adding labels that hopefully explain that no, we aren't going around chasing down and killing animals just for the fun of it. Unfortunately, sometimes people make assumptions as to what terms like "cruelty-free" means. They may assume the animal died naturally and was collected by the seller/owner. In actuality, this is sometimes the case, but sometimes the person also includes roadkill and other accidental deaths under the "cruelty-free" label. And unfortunately there are unscrupulous people who will refer to their wares as cruelty free or humane when in actuality they aren't, all so they can get sales from the people they mislead.

If you use another person or entity's label, please make very sure you're representing yourself accurately. For example, I've seen at least one seller online using the "Origin Assured" label in asserting that their animal remains only come from sources that don't use leghold or snare traps, farmed animals are not solely kept in cages, and a number of other claims. In fact, according to http://www.furcouncil.com/originassuredfur.aspx "Origin Assured" just means that the fur came from a country that has regulations governing the care of furbearing animals.

My personal choice is to go with the term "eco-friendly", for a couple of reasons. A lot of my materials for my hide and bone art are either industry scraps from making garments or other items, or they're secondhand or found in nature. In fact, on my online listings I specifically describe which materials are new and which ones are secondhand, and I also try to include as much information as I can on their origins. I also work very hard to use everything I have, down to the tiniest scraps. And I donate a portion of the proceeds from their sales to nonprofit organizations that benefit wildlife and their habitats; my self-employed status also gives me the time and schedule flexibility to volunteer with such organizations. And, of course, hide and bone will biodegrade over time, while artificial alternatives made of petroleum-based plastics won't. So by my personal standards that's eco-friendly, and as an environmentalist it's the best solution I've found in sourcing and utilizing my art materials.

Obviously not everyone will agree that stuff made from dead animals is eco-friendly, but I feel I've done my best to be transparent about what that means to me and why I use that label. I also review my reasons periodically, as well as any time I get new information about my sources, for example if someone has been deliberately misleading me. My suggestion to you is to also endeavor to be as open as you can, and understand that there are always going to be people who will disagree with what you do no matter how you label it.

Questions to Ask Yourself About Ethics

You may already have a pretty good idea of you ethical boundaries. However, these are questions that I ask myself periodically to assess whether my own ethics have changed over time, and I thought they might be useful to you as well:

- What are my boundaries regarding how an animal dies?
- Am I okay with animals that were hunted, trapped or farmed (by farmed I mean specifically farmed for hides, not food)?
- What about those whose death I am unsure of, since I may not have much, if any, information about their origins?
- Am I okay with killing an animal for food, such as a chicken or rabbit, and using its remains in my collection or art?
- Would I ever kill an animal for its remains, such as hunting a coyote for fur?
- Am I okay with new remains, as opposed to vintage or otherwise secondhand ones?
- Do I need to process the remains myself, or can I also collect those processed by other people, even if I don't know their exact origins?
- What are my thoughts on collecting the new remains of a species that is legal to hunt, but whose population may be declining, or whose status may be unknown due to a lack of information on population trends?
- How do I feel about the vintage or otherwise secondhand remains of an endangered species that may be illegal to hunt, but still legal to possess and/or buy, sell and trade?
- If a seller has knowingly traded in animal parts that are illegal or whose origins I don't agree with ethically, am I okay buying something from them that is legal and within my ethics?
- Am I okay with buying legal animal parts from countries that don't have good track records with regards to protecting endangered species and their habitats?

These questions can help you think about your ethics, whether you're figuring them out for the first time, or revisiting them. Consider them a starting point, not the beginning and end of your meditations.

Collecting Without Going Broke (Or Homeless)

There's a tendency among new Vultures to acquire any scrap of hide, bone or feather they can get their hands on. I did this when I was in my late teens, got my first job and an income of my very own. I quickly amassed a collection consisting of a motley assortment of mismatched antlers, a bag of scrap leather from a shoe shop, a couple of old fur coats, a handful of bone beads and a poorly mounted monitor lizard. Someone gave me a badly dessicated garter snake, and I scored my first deer skull at a yard sale.

Once I moved into my first apartment with a limited amount of display space, I began to narrow my collection's focus out of self-defense. For a while I was all about pelts, and there was a growing pile of wolves, badgers, foxes and other critters that threatened to replace the bed as my favorite place to snooze. But they were big and bulky and prone to moths, so I sold off or otherwise re-homed all but a few of them and switched to collecting skulls. I'm partial to professionally cleaned specimens, complete with all teeth and other bits, though I'll give a little more wiggle room to rarer specimens, like the juvenile spotted hyena missing its nasal bones or my capybara's broken-and-glued mandible.

Even though I've moved since then, there's not a lot of room for extra stuff. That means that periodically I go through my critter bits, to include both personal collections and art supplies, and pare the numbers down some. I'll talk more in Chapter 5 about display options for small spaces, but for now let's look at how to plan your collection so you're less likely to run out of space and money.

The first question you want to ask yourself is:

What do I like and why do I like it?

By this I mean what parts of Vulture Culture do you like the most? Are you partial to fur, the colors of the pelts and the soft textures? Or do you like bones, the sturdy structures of vertebrate animals? Do you find wet specimens to be the opposite of gross? Maybe you like particular groups of animals, such as canids or snakes, and want to collect the remains of any of them you can legally and ethically gather.

There's a good chance a bunch of you are going to answer my question "All of it! Because it's AWESOME!" That does make things a bit more challenging, but not impossible. Here are a few more questions to consider:

- Even if you like everything about Vulture Culture, is there anything in particular that especially catches your imagination? If you had to pick just one thing to collect, what would it be?

- What sort of aesthetic do you like? Do you want a neo-Victorian mini-cabinet with a wide variety of specimens? Or are you out to weird people out with the strangest and most eccentric things you can find? Or are you more functional, wanting a selection of skulls for comparative anatomy and drawing references?

- What sort of space do you have access to, and what can you realistically fit there? Even if you've been drooling over an Alaskan moose shoulder mount with record-breaking antlers, there might not be enough room in your tiny studio apartment (assuming you could even get it through the door in the first place!)

- What's your budget like? Would you prefer to buy a whole bunch of inexpensive specimens, or save up for one big one? Or would you rather use your money for cleaning and tanning supplies?

With regards to that last bullet point, money is a really important factor to consider no matter how much space you have. The costs of collecting, particularly if you do more buying than scavenging, can add up very quickly. Thankfully, there are ways to keep Vulture Culture from eating your rent every month.

First of all, shop around. As Vulture Culture has become more popular, there are now more suppliers than ever. Some will charge full retail price for their items, which is perfectly acceptable. However, there are also wholesalers out there, and while the quality may vary a bit more since they're offering more than just the crème de la crème, the price is usually right. Keep hunting through those secondhand sources, and keep your eyes peeled for great deals in online groups. Barter is also a potential option if you have specimens to trade.

Also, despite what I said about preferring A-quality skulls, don't be afraid of a little damage. You can still get really lovely specimens with minor imperfections that bring down the price significantly. Hides can be stitched, feathers brushed, leather conditioned. And if you're halfway decent at sculpting, you can use some epoxy putty like Apoxie Sculpt to replace missing teeth and otherwise repair skulls and bones.

Finally, it's generally cheaper overall to prepare specimens yourself than buy them completed. Hide tanning and bone cleaning are messy and time-consuming, but ultimately you'll spend less, especially as you prepare more remains for your collection. If this appeals to you, you're going to love Chapter 4. But first, let's talk about what happens when that nifty dead thing you just brought home doesn't have a label on it.

Chapter 3: Identifying Hides, Bones and More

If you tend to buy your specimens new, you're almost always going to know what species they came from, with the occasional exception of a seller who really has no idea what they're doing. Secondhand shops have a tendency to mislabel taxidermy and other items, and bones you find in the woods generally don't come with handy-dandy ID tags. Over time you'll develop an eye for certain marks, shapes and colors that signify different species' remains. Until then, consider this a starting point if you're new to Vulture Culture.

Why is it so important to be able to identify the specimens you have? One big one, of course, is legality. You want to be able to tell the difference between a wild turkey feather and a great horned owl feather if you're in the US, because the former is legal to have but the latter isn't. The same goes for other easily confused specimens like jaguar and cougar skulls, or any of the remains of a legal English house sparrow versus those of all the native sparrow species.[14]

Moreover, if you want to do any sort of selling or trading in Vulture Culture you'd best know what it is you have, not just for legal purposes but also to gain a good reputation. If you keep mislabeling specimens people won't trust what you have. And even if you don't sell anything, it's just nice to know what it is you actually have. You don't have to go through the process of painstakingly labeling everything, but you should be able to identify anything you have that someone asks about.

There are three main steps that I take to suss out a new specimen if I don't yet know what animal it came from:

Compare to known species: Over the years I have seen many a dead thing, and while my knowledge is far from encyclopedic I have a pretty good mental database. So if I have a new, unidentified hide or bone or whatnot the first thing I do is figure out whether it resembles anything I recognize. Sometimes the resemblance is close enough that I can pretty safely say they're one and the same. Other times I may have to physically compare my new specimen with its lookalike to see if there are any definite differences.

What am I looking for? For hides, the first thing I get is an overall impression—the shape and size of the hide can often be enough information. I also examine the color and pattern of the fur or leather, seeing whether there are

---

14 Even when alive, sparrows can be notoriously hard to tell apart, particularly the females. Many tend to be a plain, drab brown with few markings, and are also similar in size and coloration to certain female finches and other small songbirds. There are, in fact, so many of these little mud-colored chirps and they can be so difficult for even experienced birders to tell apart that they've collectively earned the informal title "Little Brown Jobs." "Oh, hey, there goes another LBJ. Any idea what it was?" "Not a damned clue."

different layers in the fur as with canids, or whether it's a single flat layer of hair as on a cow's summer coat. I also handle it and get a feel for its weight, thickness and texture. There are some hides that have a very distinctive feel; deerskin leather, for example, is soft and supple but stretchy, and its scent to me is uniquely sweet.

For feathers, U.S. Fish and Wildlife has a fantastic resource, the Feather Atlas. Located at http://www.fws.gov/lab/featheratlas/idfeather.php, it's an excellent tool that allows you to compare a feather you have to those in their database. Right now it only includes flight feathers from the wings and tail, which are the largest and most distinct. Still, it's a great way to learn which feathers are from legal species and which aren't.

What if you don't know very many animal parts by sight yet? Start researching! Skulls Unlimited's website at http://www.skullsunlimited.com has an impressive array of both real and replica skulls, and Bone Clones at http://boneclones.com has even more museum-quality replicas. Identifying hides often takes both seeing and touching them, but you can brush up on visual identification by browsing the Tanned Furs section at http://www.hideandfur.com. Both that site and Bone Clones have decent collections of smaller bits and pieces like teeth and claws (replicas in the case of Bone Clones) if you want some practice identifying those. And since these sites are rather nice about providing this free research material, I highly recommend making a purchase from them as a thank you.

Taxidermy shows are also great places to get a good look at a wide variety of animal parts, especially hides and antlers, though the vendor room may also be full of skulls and the like. Wander the exhibits and take a close look at what's on display. Generally speaking you won't be able to touch the specimens, but vendor rooms are usually more hands-on. You can also usually take photos which you can then add to your personal reference library, though again some people may not want their work photographed. When in doubt, ask before touching something or taking a picture of it.

Analyze physical characteristics: Animals evolved different traits depending on which adaptations were best at helping them pass their genes down to the next generation. These traits can tell us a lot about each species' environment, their response to it, and most importantly to us—their identity.

As before, get a good overall sense of the size and shape of the hide, bone or other remains in question. This gives you an idea of how large the animal was when alive; you know that a narrow hide about twenty inches long will be from a smaller creature, not a deer or wolf. If it's a complete hide, look at how long the legs are in comparison to the rest of the body (note that the legs can shrink a little bit during the tanning process.)

Next, check out details. This long, skinny hide came from a long, skinny animal, and this particular hide is brown with a somewhat soft, fluffy tail, and an overall sleek appearance. The face isn't too long, and it has tiny ears. If you guess that it's some kind of weasel, you're probably right! After that, you'll need to compare it to other hides to ferret out its exact identity.

You can use this same attention to detail to explore other remains and narrow down possible identities. Later in this chapter I've included a skull identification key, in part to give you a little more practice in the art of analyzing physical traits.

Ask for help: One of the great things about Vulture Culture is that Vultures LOVE playing "What's that skull/hide/claw?" So all you have to do is ask in the right places, and you're likely to get answers. Chapter Y goes into more detail as to where Vultures like to flock online and in person, but even if you know only a few you can still email, call or ask them in person.

If you're asking online, you want to get as good a set of photos of the piece in question as you can, and make sure at least one is taken with a ruler to show scale. Have good lighting available; I like natural light myself, preferably outdoors on an overcast day. Take the photos with whatever camera you have at your disposal, even if all you have is your phone. Use an image editing program to increase the contrast or sharpen the picture to get a bit more detail, either Photoshop or a freeware program like GIMP.

Other important information to share is where you got the specimen in the first place, and where you are now. This isn't just so people can narrow down the geographic origin of the specimen, but also so others can give you a heads-up if it turns out it's something that's not legal to possess. If it does turn out to be illegal, put it back out in the woods or return it to the seller ASAP.

The more specimens you can identify, the easier it will be for you to look at something new, see that it's identical or very close to something you already know, and be able to say "Yes, that's a coyote hide" or "this is definitely a domestic rabbit skull." Be patient with yourself; it generally takes years to build up a lot of knowledge, and even those of us who have been at this for decades still get stumped sometimes.

The rest of this chapter is going to be dedicated to more specific modes of identification, along with some photos of hides, skulls and feathers. Since I can't include the identities of every single bone, hide or feather out there, I'm going to stick to some of the most common North American animals, primarily mammals. I realize this is pretty limiting, especially for those of you not in this area. However, the concepts are the same—**compare to known species, analyze physical characteristics, ask for help**. You're always welcome to email me at lupa.greenwolf@gmail.com if you have something you haven't been able to positively identify. I can't 100% guarantee I'll have the answer, but I'll do my best, up to and including consulting other people who may have a better idea of what you've got.

Some Common Hides (With and Without Fur)

Let's start with leather and fur. Most leather is from domestic animals like pigs, cows and goats, and is made when the hair is removed from a hide and it is then preserved, either through drying as in rawhide, or through tanning. Rawhide is commonly used to make lamp shades, drums and rattles, though it is also popular as a chew toy for dogs.[15] It's stiff and generally fairly thin, though it can be made supple again by soaking it in water. Leather is a lot more versatile, being used for clothing, saddles, upholstery, and much more.

If you hear or read about leather being measured by weight, it actually means thickness. 1/64" equals one ounce of leather weight, so a hide that is 10/64" thick has a weight of ten ounces. Most hides have a variable thickness, usually thinner around the edges, so many hides may be tagged in a weight range, such as "4 to 5 ounces." To complicate matters more, leather can be thinned through a process called skiving or splitting. Suede leather is made from the underside of a leather hide when it's split, which is how it gets its soft texture on both sides compared to the leather made from the top of the hide or an unsplit hide, which has a smooth side and a soft, suede-like side.

It's really hard to convey how to identify different types of leather through pictures, so I'm going to do my best to try to describe how I tell the difference between some of the more common ones. Please be aware these are my subjective experiences, and it's best to handle the hides themselves to get a feel for how you can differentiate between different species. Also, most leather garments just say "100% leather" on the tag, and don't tell you what species, but most of the time it's going to be one of the first three of these.

Cow: This is one of the heavier hides on the market; it's the leather you see on biker jackets and chaps, and thicker "tooling" weight cowhide can be used for purses, saddles, belts and the like. Tooling leather tends to be stiff, though not as much as rawhide, and its surface can be carved, shaped and dyed with various tools. Lighter cowhide ends up in heavier garments, and it has a distinctive durable feel to it, similar in weight to canvas tarps. It can be anywhere from a couple of ounces for thin-split suede, to ten or more ounces for tooling leather. Calf hides are also sold; these tend to be thinner, of course, and similar to deer or lamb but not quite as supple.

Pig: This is a lighter, stretchier hide, often sold as suede splits. You see pigskin suede a lot in the fashion industry as jackets, vests and skirts. Pig leather has a distinctive appearance due to the many tiny holes all over it; these are from the hair follicles. It tends to only be a few ounces; 1.5 to 3 is most common.

Lamb: Even softer and more supple than pigskin, lamb can be quite delicate, like a fine fabric, and generally has a very smooth texture. 1 to 1.5 ounces is a common weight. Lambskin is frequently sold dyed, coated and otherwise altered.

---

15 I really don't recommend rawhide for dogs. Not only is it often bleached with chemicals that are bad for them to ingest, but if they swallow a large enough piece it can get stuck in their esophagus. Even if it makes it further into the digestive system, it can swell up with digestive juices and cause a painfully fatal blockage.

Deer: This is my favorite leather. It's generally only 2 to 4 ounces, but is surprisingly strong and stretchy for its weight, making it a great option for lacing and clothing. It's quite soft, though not as much as lambskin. It also has a distinctive, almost sweet scent to it.

Elk: Think deer, only an ounce or two heavier on average. Whole elk hides are huge, but even scraps feel heavier than their deer counterparts on average. The thickness means that it doesn't fold quite as neatly or delicately as deer; over time you'll get a sense for the physical heft of elk that deer just doesn't have.

If you happen to have a Tandy Leather branch or other leather store nearby, I highly recommend spending some time familiarizing yourself with their stock. You might even ask if they'll sell you some scraps from each species they carry so that you can have your own sort of reference guide! Do be aware that when leather is color-treated, given additional texture, or coated with sealants that it can change the appearance and feel, so make sure you look at those altered hides as well as natural ones.

I also want to briefly mention reptile leather and ostrich leather. You can get tanned alligator hides, as well as the skins of several species of snake and lizard. Unless dyed, these look very much like the live animal, including texture and color (though the colors may be muted in some cases.) Ostrich leather is similar to pigskin, down to the marks from where the feathers once grew. However, the hair pores on pigskin are smaller and often come in clusters of three, while ostrich feather pores are singular and larger.

Fur is a much more complicated matter. There are literally hundreds of species whose hides are in legal circulation, and some of them look like almost nothing else in this world. I'm going to show you some of the more common furs you may find floating around Vulture Culture, especially in North America.

First, a bit of terminology. When a pelt is removed from an animal, a

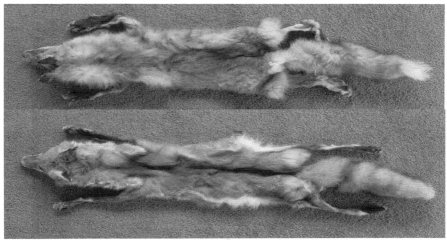

professional skinner will keep the belly hide intact and pull the entire hide off over the head starting with the tail end. It's kind of like taking a sweater off inside-out, only involving the animal's entire skin, often including feet and all. This is called a cased or barrel skin. If the belly of the skin has been cut open partway or all the way so the hide lays flat like a rug, this is called an open skin or abdominal cut. The photo above shows a cased skin at the top, an open skin at the bottom; notice how you can see the white of the belly on the edges of the open skin. If a pelt only includes the head, neck and shoulders, this is known as a cape; these are usually used to make taxidermy mounts, and if they are split the cut is made down the back of the neck so the stitching when it's sewn back together is easier to hide.

Speaking of taxidermy, most hides are not of sufficient quality to use in a traditional taxidermy mount. The hide needs to be exceptionally complete, to include all feet and claws/hooves, lower jaw, lips, eyelids, nostrils, genital hide, etc. and lacking any large holes or other damage. The ears, lips, nostrils and eyes all need to be "turned", meaning all excess flesh and meat needs to be removed, to include on the inside. The tan needs to be of a very high quality as the hide will need to be stretched tightly over a plastic form without breaking. A hide can be either wet tanned or dry tanned; wet tanned simply means that after the tanning process is done the hide is not allowed to dry out, but is oiled to prevent it from drying and then either used immediately or frozen. A dry tanned hide is, as the name suggests, allowed to dry out and then can be rehydrated before use. Each taxidermist has their own preference for wet or dry tans (or both!)

While I can't show you picture of every single hide, I wanted to give a few examples to get newcomers started. I apologize for these photos being in black and white; if I'd had this book printed in color it would have been significantly more expensive. To make up for it, I have color versions of the photos of the hides and skulls available on my website, http://www.vultureculture101.com. Also, a special thanks to Moscow Hide and Fur in Moscow, ID for allowing me to come to their warehouse and take photographs of many of the hides and skulls in this chapter.

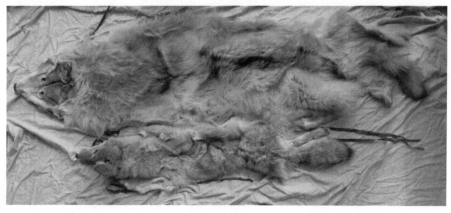

This picture shows a gray wolf above, and a coyote below (the straps are because these are wearable headdresses I've made.) Wolf and coyote are often confused, especially if you have a particularly thinly furred smaller wolf and/or a very plush large coyote. However, wolf fur generally has much longer guard hairs, especially over the shoulders, and is more likely to come in colors other than some variant of brown. Wolves are also usually much larger; a wolf is going to be in the sixty to eighty inch range, generally speaking, while a coyote will usually be forty-five to sixty inches in length.

Most coyotes will be some shade of brown; however, they can also come in a reddish color similar to a red fox though with more white on the face and flanks, and no white tip on the tail. A melanistic coyote is very dark brown to almost black; here's a picture of a regular coyote next to a melanistic one:

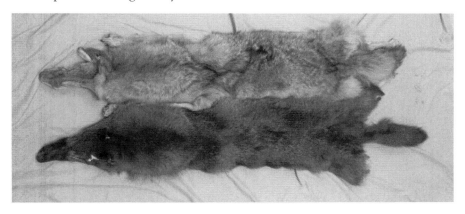

Along with coyotes, you're likely to see a lot of fox hides for sale. Here are four you may encounter:

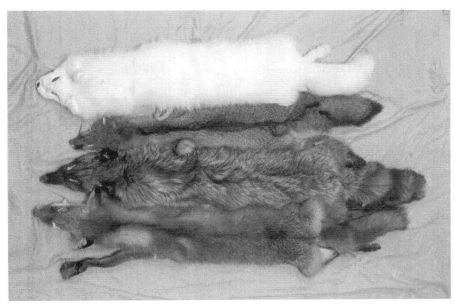

At the bottom you can see the fur of a red fox; this species has dozens of color variants, mostly found in ranched rather than wild animals. Just above it is a silver fox, which is black with a scattering of white. The only colors that tend to occur in nature are red, silver, cross (like silver but with red patches here and there) and very, very rarely leucistic (with some white patches or all white) or albino (white with pink eyes and skin) foxes. You're not going to find a pearl cross marble fox in the wild, unless it's an escapee from a ranch that somehow managed to survive, which almost never happens.

The gray fox is an older, smaller, more primitive species than either the red or Arctic fox. They have a distinctive gray grizzled coloration with black marks on the muzzle and tail, and red patches on the ears, underside, legs and tail. The second hide from the top is from a gray fox.

The top hide is an Arctic fox. Arctic fox furs are usually white, as that's their color phase in winter when hunting and trapping generally happens. However, they can also be brown or gray in summer, and sometimes summer hides find their way into the fur trade. On very rare occasion you may find the winter hide of a brown or gray Arctic fox; this is from a coastal population in Canada that doesn't turn white.

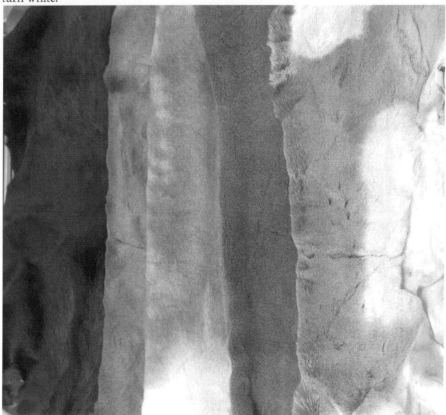

These five hides are all from various ungulates Each is covered in many smooth hairs which, upon inspection, are actually hollow; these hollow guard hairs capture warmth better and help insulate the animal during cold weather. However, the four on the left are in the deer family—moose, elk, caribou, and whitetail deer. The one on the right is from a pronghorn antelope, a distant relative that is the last member of its family, *Antilocapridae*. It's actually more closely related to giraffes than to deer!

If you were to handle these hides you might notice that the caribou feels a little softer and more plush than the others. The moose has dark, very long hairs; the elk and deer feel similar, though the hairs on the deer hide are thinner and more delicate. The pronghorn's hairs are especially prone to sticking out if you fold the hide, All five of these animals have distinctive colorations, so after you've handled each in real life you should be able to distinguish them.

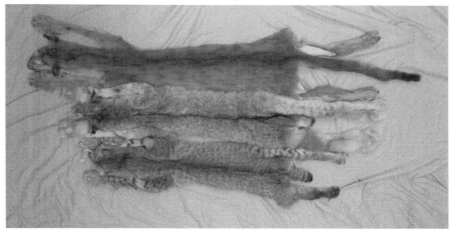

These are the three most common legally traded feline hides in Vulture Culture. The first thing you'll notice when you handle them is that they are all quite soft to the touch. The bobcat hide (bottom) is by far the most common; this animal lives in much of the United States and Mexico, and a small portion of Canada. The upper part of the hide is golden brown with lots of black spots, while the underbelly is white with more spots along the edges. The tail has black stripes, though in some animals these may be minimal at best.

Canadian lynx (center) are generally larger than their bobcat cousins; their coat is a little more silver toned. They also often have subtle spots in their coat. The end of the tail is black, like it's been dipped in paint. The tufts on the ears and the cheek fur are longer as well.

The mountain lion (top) has an overall more even coloration, being a reddish tawny brown without spots, and a brown or black tip to the tail. You'll notice that the tail is very long, especially in comparison to the short tails of the bobcat and lynx.

Here we have a quintet of mustelids. The bottom two are the same species, but the one on the bottom is a wild mink, and the other is a captive-bred mink. Captive mink come in many more colors, and the hides are frequently color-treated.

A marten (center) looks a great deal like a wild mink, but it's more robust and has longer legs. And the fisher (second from the top) is larger still. The otter (top) is the largest of all the commonly found mustelid hides, and has a uniquely sleek coat with very dense, short hairs that help insulate it and keep its skin dry even when immersed in water. All three of these tend to be medium to dark brown, though oddball variants do occur in the wild occasionally.

The largest mustelid in the world is the wolverine. The photo to the left

shows a headdress I made from a wolverine hide a number of years ago. Note the unique pale tan pattern on the dark brown body; this is something every wolverine will have, other than extreme leucistic or albino individuals which are exceedingly rare.

This is an assortment of medium-sized hides that you'll commonly find in Vulture Culture. The skunk (bottom) is unmistakable with its black and white coloration; striped skunks like this one are most common, though the rarer spotted skunk is protected in some areas as its populations have been in sharp decline. The opossum (center), North America's only marsupial, has a cream-colored undercoat with somewhat scraggly gray or gray-brown guard hairs and a hairless tail. The raccoon (top) is one of the most common hides of all, and is generally brown with lighter guard hairs, and a distinctive mask on the face and striped tail.

In spite of being driven nearly extinct in the 19th century, American beavers have rebounded and their reddish-brown hides with coppery guard hairs are an easy find on the fur market. The photo on the next page shows two beaver hides. One is cased, and the other is split and stretched into a round shape with the tail and limbs removed. The latter shape was used to make transporting the hides easier in the 1800s when beaver fur hats were trendy and beaver pelts were in very high demand.

The muskrat (not pictured) looks similar, but is less red in color, has shorter hairs, and the whole pelt is much smaller. Beaver fur can also be much darker, nearly black. If you ever see whole hides, muskrats in addition to being smaller also have a long, whip-like tail, while the beaver's is characteristically flat.

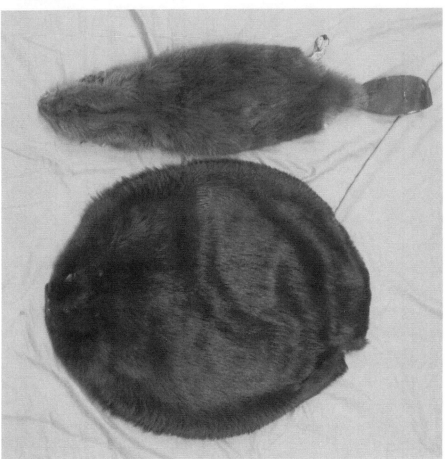

There are many, many more pelts out there that you're likely to run across as you continue your journey as a Vulture, but I hope you find this to be a good starting point! Again, when in doubt, compare to known species, analyze physical characteristics, and ask for help.

Bones in the Body

With the exception of fish, all vertebrate animals are tetrapods—literally meaning "four feet". And we more or less have the same skeletal structure; it just evolves different details in different species. For example, almost all mammals (save for manatees and sloths) have seven neck vertebrae. And while horses only have one large middle toe left[16], their ancestors had five digits on all four feet, just like us.

---

16 You can occasionally find horses that have one or two vestigial toes on their lower legs, sometimes with complete, if small, hooves. And little calloused areas, called chestnuts or ergots depending on where they are on the leg, are also believed to be tiny remnants of the second and fourth toes.

Because every tetrapod has basically the same general array of bones, after a while you can get an idea of what part of the body a given bone came from, even if you don't know its species. For example, a long, slender, straight bone with knobs on the ends is going to be from an animal's limb, while a cylindrical or boxy bone with a hole in the center and various tabs and other outgrowths (known as "processes") is a vertebra. After a while you'll be able to pick up a random bone and tell whether it was a rib or leg bone, or what part of the foot it likely came from, or whether it's complete or not.

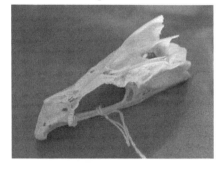

Not all bones are as easy to identify. For example, the picture to the left shows a drum fish skull. It's commonly mistaken for a bird breast bone or a partial tetrapod skull before.

What follows is a list of general bone categories and how to identify them.

Skull: This is the most obvious one in most cases. You can see the eyes and nostrils, and if you have both the main portion of the skull and the jawbone you can see how the entire thing fit together, especially if all the teeth are present. Here's a diagram showing the basic bones of a skull, modeled by a black bear:

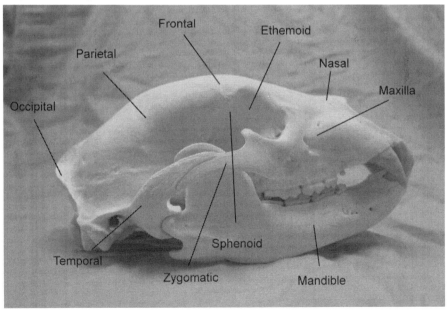

Mammals are unique in having several different types of teeth; while generally speaking reptiles, fish and some amphibians only have one sort of tooth, usually a simple spike or peg. (Birds, of course, have none, unless we're messing with their genes in the lab and bringing back their dinosaur heritage.) Most mammals will have some or all of these: incisors (the teeth in the very front of the mouth), canines (the big sharp teeth), premolars (the teeth just behind the canines) and molars (the teeth at the very back of the mouth). Here's our black bear again, showing off her teeth:

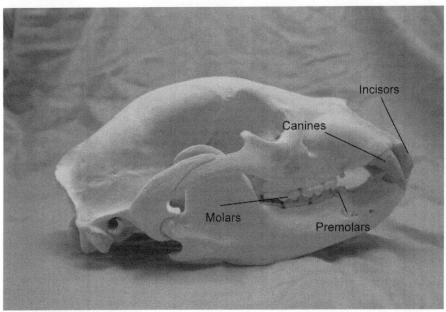

I'll talk more about using teeth and other features to identify skulls in just a moment. For now, let's look at some of the other bones of the vertebrate body you might find or purchase.

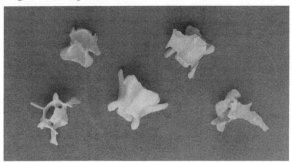

Vertebrae: Generally speaking, tetrapods have several different specialized types of vertebrae: cervical (neck), dorsal (back, including thoracic ones with ribs attached and lumbar ones without ribs), sacral (just above tail) and caudal (tail). Birds and some reptiles (most notably dinosaurs) are unique in that a few of their thoracic, all lumbar and sacral, and a few caudal vertebrae are all joined together in a structure known as the synsacrum. This unusual looking bone structure is sometimes mistaken for a skull.

Scapulae: These are the shoulder blades, and if all goes well your average animal has two of them. They're easy to identify, a flat, round or triangular piece of bone with a sort of cupped tip, almost like the mouth of a bottle, and a ridge of bone on the backside. This fox scapula has more rounded edges; some mammals, like deer, have more triangle-shaped scapulae.

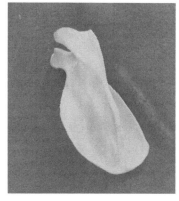

Pelvis: Another pretty unique bone, in mammals it almost looks like a face (and in fact some artists have made large ungulate hipbones into amazing masks.) Split in half, mammal pelvis halves look almost like a couple of magnifying glasses. Bird pelvises are fused into the synsacrum; reptile and amphibian pelvises vary widely according to the type of animal, and I recommend wandering around Google a while searching for things like "frog pelvis" and "lizard hips". (Make sure you have the mature content filter on just in case.) Snake pelvises are about as common as hen's teeth. Some more primitive species, like some constrictors, have remnant leg bones that terminate in spurs next to the vent, but that's all that's left. The next two pictures show a split pelvis, and a whole one I made into a planter.

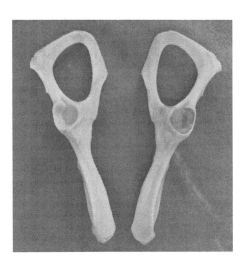

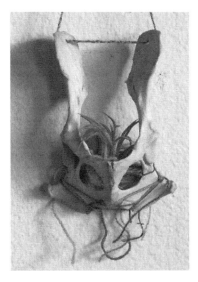

Ribs: These are usually one of the easier bones to identify; they're a curved arch with a nub at the top that looks a little like a horse's head. There are species that have very straight, flat ribs; some of the shorter floating ribs of bison are a good example. In birds, the ribcage includes a very large breastbone. The muscles of the wings are anchored to it and help power flight. The picture on the left below shows two small mammal ribs and a snake rib on the right.

 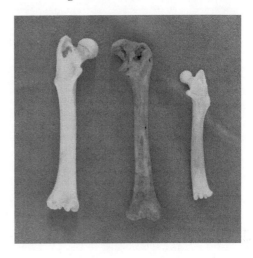

Leg bones: You'd think these would be as easy as ribs to identify, but you'd be wrong, at least initially. There are actually six type of leg bone: femur (upper back leg), tibia and fibula (lower back leg, fibula is generally very thin), humerus (upper front leg/arm/wing), radius and ulna (lower front leg/arm/wing.) At first glance, especially in smaller animals, these may look confusingly similar, but with a little practice you'll get to know what makes them unique—for example the large head of the femur, the fibula's skinny middle, and the C-shaped end of the ulna. Keep in mind, of course, that the same bone may be shaped somewhat differently in different species, especially once you start comparing animals from different classes. I've included a few examples of small mammal femurs in the photo above on the right, but these should not be your only source for identifying long bones; Google is your friend there.

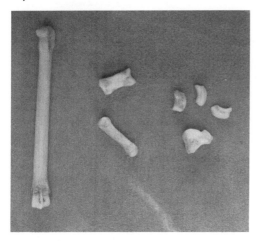

Foot bones: This is another set of bones that people can sometimes find confusing. Generally speaking, all animal feet (which includes our hands) have carpals, metacarpals, and phalanges. On your own hand, the carpals are the bones in the wrist, the metacarpals are the bones in the palm of the hand, and the phalanges are the fingers. Many mammals walk on their toes, comparatively speaking; if you look at a dog's forelegs, the dog's pastern is analogous to our wrist as both are made of carpal bones, and they walk on their phalanges. In some animals the limbs are heavily modified, as in bird wings or horse's lower legs, where the bones have been fused together. The photo above shows a coyote metacarpal on the left, two phalanges in the center, and a few carpals on the right.

Miscellaneous: The rest of the bones of the body tend to be small; tetrapod ankles and wrists tend to include the aforementioned carpals, and the patella (kneecap) is also not especially large compared to other bones. Small pieces of the skull may also break off. Many of these bones are hard to identify unless you have had a lot of practice in articulating skeletons, and sometimes may be mistaken for bone fragments.

Skull Identification

Skulls are the bones most prized by many Vultures. They're also particularly easy to analyze for identification because they're so distinct. After that first overall look, the next thing I check is the teeth. The teeth of reptiles are generally all alike—crocodiles have cone-shaped teeth while a king snake's teeth all look like a bunch of little hooks. The one really big reptilian exception is certain snakes that have longer, usually curved fangs, but even these are merely longer versions of the rest of the teeth.

Mammals have more specialized teeth—some combination of incisors, canines, premolars and molars can be found in almost all mammals save for echidnas and adult platypus. These teeth and their shapes can tell you whether a mammal is a carnivore, herbivore or omnivore. Teeth are incredibly important to skull identification; even if all you have is a few teeth, you can tell a lot about the animal's diet and behavior. In the photos below you can compare the jaws of a red fox (left), opossum (center) and Chinese water deer (right).

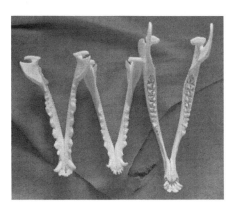 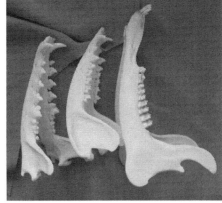

On the left, the red fox jaw has very prominent, sharp canine teeth; the incisors are small and good at snipping away skin and feathers to get to the prey's meat. The premolars and molars are also sharp and can be used to shear off larger chunks of meat, like a pair of scissors. The water deer on the right is an herbivore jaw; notice how the incisors are larger and flatter, good for cropping plant matter. There are no canine teeth as they aren't necessary. The molars are flat so the animal can chew the plant matter and break down the fibers for easier digestion. In the center, the omnivore opossum jaw has canines and sharp incisors, but flatter molars than the carnivore has, showing that this animal can eat a variety of both plant and animal based foods. Your own teeth are a great example of omnivore dentition!

You want to ideally be able to look at the entire set of teeth, if possible, to figure out what an animal eats. If all you had was our black bear's canine tooth, you might think she was a ferocious carnivore. In fact she is an omnivore, and the majority of her diet is plant based. Her molars are flatter than those of true carnivores like wolves.

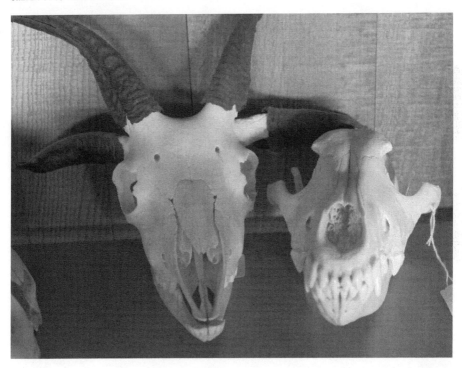

Another thing to notice is the placement of the eyes. Generally speaking, herbivorous mammals will have their eyes on the sides of the head, like the Jacob's sheep on the left. Next to him, the gray wolf has eyes that are more forward-facing. Wolves and other predators need to be able to focus on what's ahead of them, especially when hunting. Herbivores, on the other hand, need as wide a range of vision as possible so they can keep an eye out for predators while they're grazing.

Obviously there are a lot of features of skulls I've not covered here. For example, both lagomorphs (rabbits and their kin) and rodents have two pairs of long, ever-growing incisors, but how do you tell which family the skull you have is in? What about bird, reptile, amphibian and fish skulls?

This book is just a very brief introduction. Appendix I has some book recommendations that can help you identify all sorts of skulls, to include in-depth identification keys. There's not really space to include an entire identification key here, especially when you have skulls as diverse as white-tailed deer, American bullfrog, and Asian carp. Instead, I'm going to include photos of some of the most commonly found North American skulls so you have a quick reference. Please understand that there are many, many more skulls out there, and this is just a way to help you get to know what you're most likely to run across out in the field or in a shop. Note that measurements are averages and skulls of the same species may vary in size.

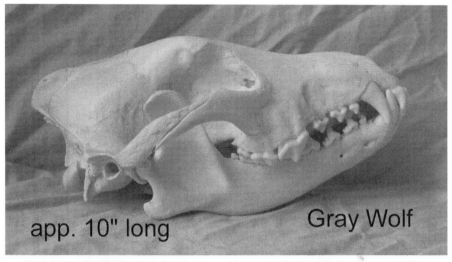

app. 10" long — Gray Wolf

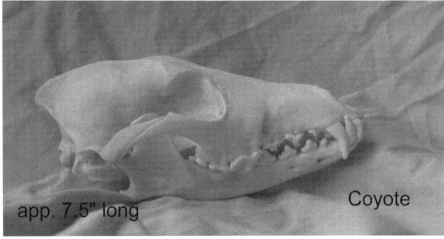

app. 7.5" long — Coyote

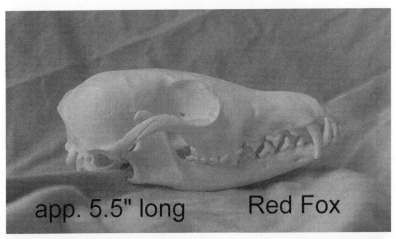

app. 5.5" long   Red Fox

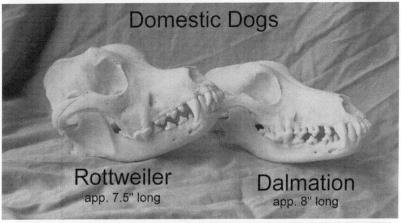

Domestic Dogs

Rottweiler
app. 7.5" long

Dalmation
app. 8" long

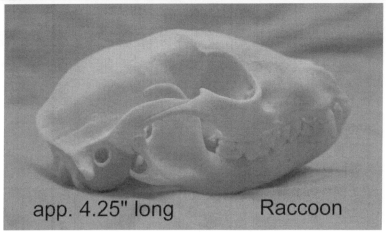

app. 4.25" long   Raccoon

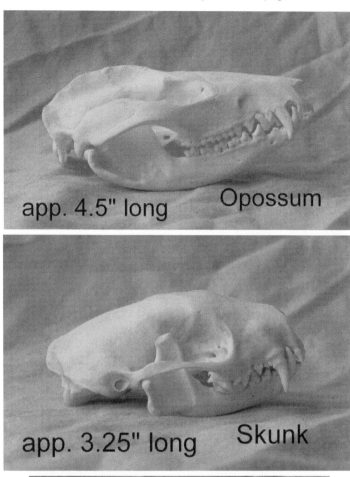

app. 4.5" long   Opossum

app. 3.25" long   Skunk

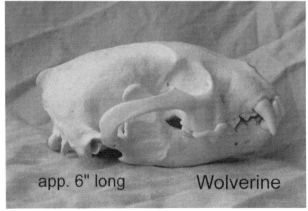

app. 6" long   Wolverine

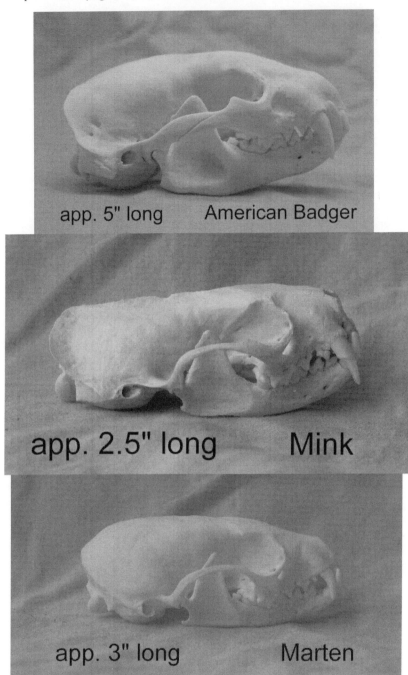

app. 5" long   American Badger

app. 2.5" long   Mink

app. 3" long   Marten

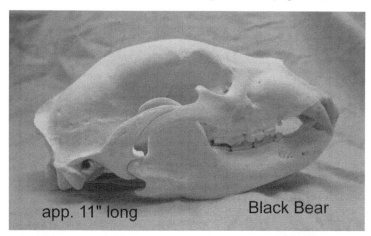

app. 11" long   Black Bear

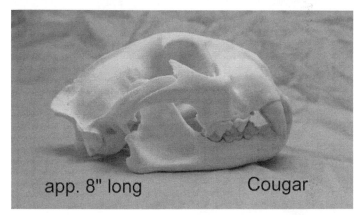

app. 8" long   Cougar

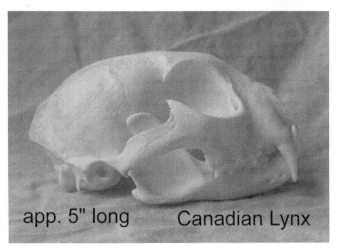

app. 5" long   Canadian Lynx

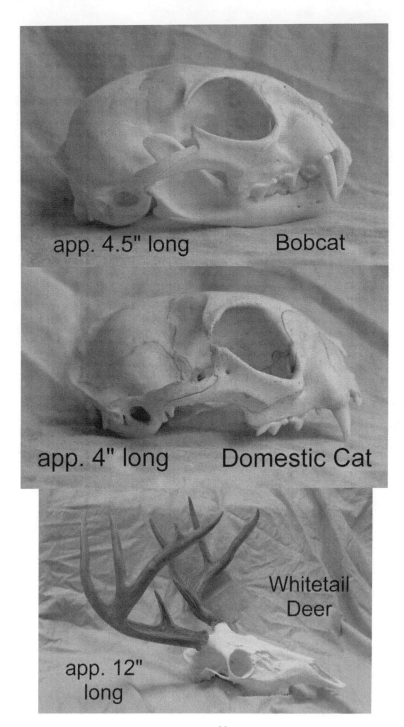

app. 4.5" long   Bobcat

app. 4" long   Domestic Cat

Whitetail
Deer

app. 12"
long

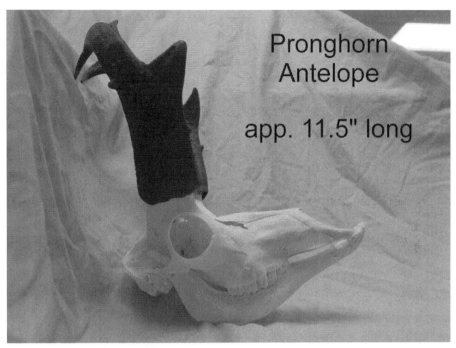

Pronghorn
Antelope

app. 11.5" long

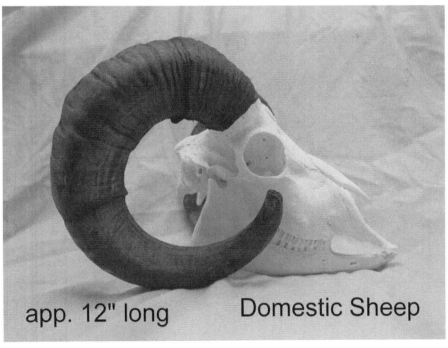

app. 12" long   Domestic Sheep

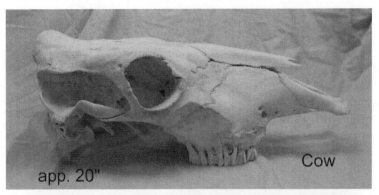

app. 20" Cow

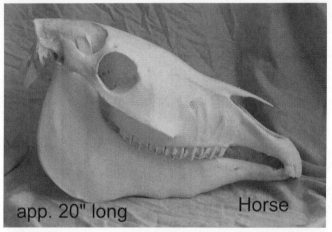

app. 20" long Horse

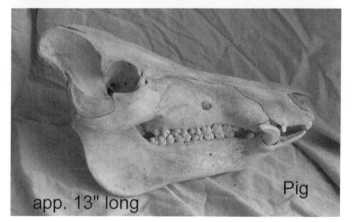

app. 13" long Pig

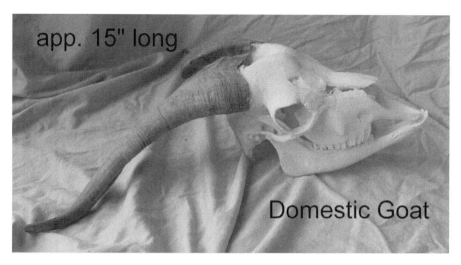

app. 15" long

Domestic Goat

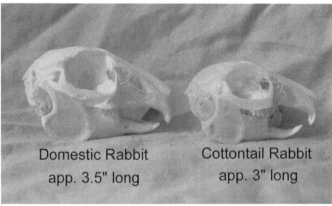

Domestic Rabbit
app. 3.5" long

Cottontail Rabbit
app. 3" long

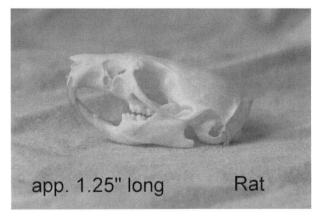

app. 1.25" long   Rat

app. 4" long   Armadillo

Okay, maybe you're less likely to run into an armadillo skull than a coyote skull. But I wanted you to know what the armadillo looks like, just because it's really cool!

Frequently Confused Skulls

No, I'm not talking about the cranial contents of certain politicians. Rather, I want to discuss a few groups of skulls that are often mistaken for each other, and how to tell them apart.

Dog vs. Wolf

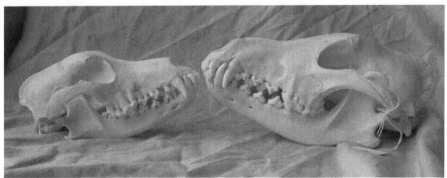

On the right we have a gray wolf skull, and on the left is the dalmation from a few pages ago. The dalmation has a more prominent stop—the curve of the forehead from the top of the skull to the muzzle. Notice how smooth the wolf's stop is, with a much flatter topline; almost all dog skulls will have a more dramatic stop than this. The canine teeth are also more substantial on the wolf skull. And the sagittal crest at the back of the wolf skull is larger than most dogs' as it anchors the substantial muscles used to deliver a powerful bite to prey, though the Rottweiler we saw earlier in this chapter has a respectable crest, as do other breeds with strong jaws.

Wolf vs. Coyote

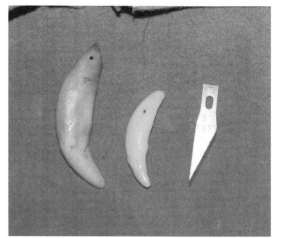

Another skull commonly mistaken for a wolf is the coyote. They have a very similar shape, but the coyote is much smaller and, even in larger males, is more delicate than the wolf skull. While you can compare the two to each other in their pictures earlier in this chapter, here I wanted to highlight the canine teeth in particular since they are the most common case of mistaken identity. The wolf tooth on the left is substantially larger than the coyote tooth on the right. Coyote canine teeth are frequently sold (accidentally or intentionally) as wolf canines. Generally speaking if the tooth is under an inch and a half long, and especially if it costs you under $20, it is almost certainly a coyote tooth.

Dog vs. Coyote

Large dog skulls and coyote skulls are similar in size, and from the side they may look particularly similar. However, the bird's-eye view shows that the dog skull on the left is more robust than the more slender coyote skull on the right.

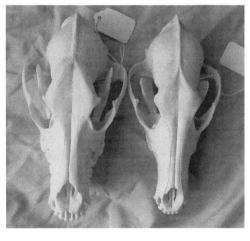

Deer vs. Sheep vs. Goat

Wandering around any field in rural America and you may run across an ungulate skull with no upper incisors. It's likely to be one of three animals: deer, sheep or goat. If it has no horns or antlers, it may be a bit tougher to identify. The deer skull's length to width ratio is much greater than that of most sheep and goats, giving it a longer, sleeker appearance, and the muzzle tapers to a finer point, especially in fawns. Sheep and goats have a generally boxier overall shape. The sheep's muzzle will usually have a convex (curving upward) shape, while the goat's is concave (curving downward). Obviously there will be individual differences depending on the species of deer or breed of sheep or goat, but these are good general guidelines.

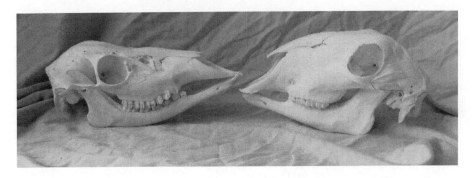

Pardon the lack of a goat skull; I tried sourcing a hornless one while working on this book and came up empty-handed. The skull on the left is a whitetail doe, and the one on the right is a female sheep. Look at the skull of the male goat earlier in this chapter to get an idea of that species' shape. Also please note that the sheep is missing her incisors in this photo.

Lynx vs. Bobcat vs. Domestic Cat

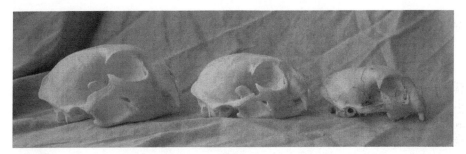

Here size also matters, though there will always be overlap. The lynx (left) is the largest of the three, with especially long canine teeth. The bobcat (center) is similar to the domestic cat (right), but is a more robust skull, while most domestic cat skulls will be smaller and more finely made than their wild cousins.

Domestic cats also have a wider range of shapes and peculiar features, though less so than domestic dogs. Notice how large the eye socket of the domestic cat is in relation to its skull, compared to the other skulls. This was a Siamese cat, bred for wider cheekbones than some other cat breeds. If I had a brachycephalic (flat-faced) Persian cat skull to compare to these two, the eye sockets would be even larger, and the muzzle almost nonexistent.

Finally, note that the cougar skull a few pages back is much larger than any of these; once you've seen one in person you won't mistake it for a bobcat for sure!

Cow vs. Horse

I have lost track of the number of times I've seen cow skulls for sale on eBay mislabeled as horse. Yes, they are both large herbivorous creatures with eyes on the sides of their heads. But the horse is a longer, narrower skull overall, and more importantly it has upper incisors whereas the cow lacks them entirely. The cow also has a more pronounced sagittal crest than the horse.

Raccoon vs. Opossum vs. Red Fox

There's a running joke on certain Vulture Culture-oriented Facebook groups that if someone posts a skull—any skull—for identification, the stock answer is "It's always a raccoon skull". That's because 90% of the time that's exactly what's been posted. These adaptable critters are all over the place and breed prolifically, so their skulls are a common find even in urban areas. The raccoon's skull (left) is a relatively nondescript looking omnivorous mammal skull with little to no sagittal crest, a small but block muzzle, and a rounded stop. The opossum (center) looks nearly alien in comparison with a very low-profile cranium and a more pronounced sagittal crest, deeply set eyes nearer the top of the skull, and long, pointier canines. And the red fox (right) has a much longer muzzle than the raccoon, but with larger eyes and a more pronounced stop than the opossum.

I know, I know—you want there to be more! I wish I could include pictures of all the skulls you might see, with more non-mammals and all. Unfortunately, since this is a general Vulture Culture guide and not a book dedicated specifically to skull identification, I could only include what a beginner might find most useful. Again, I direct you to Appendix I for further reading.

A Note on Fossilized Bones

Fossilized bones are, physically speaking, stone, and so they don't fall under the same legalities as *bone* bones. So it is possible to acquire certain animal parts as fossils that you wouldn't be able to have as fresh bone, such as whale teeth or seal phalanges. However, you want to be 100% sure that the fossil you have is authentic rather than a newer bone disguised as a fossil with paint and other substances. One quick way to tell whether your specimen is bone or stone is to heat up a needle and then poke the specimen with it. Stone will largely be unaffected, while it will burn the bone, causing a bit of smoke to rise. You can also use a very fine (1/16") drill bit on a Dremel or other multitool, or scrape it with an Exacto blade. Obviously this necessitates causing a small amount of damage to the specimen, but it's worth it to have a clear answer on its makeup.

There are also laws specifically regarding collecting fossils, and they vary from country to country. In the U.S., for example, it's illegal to collect the fossils of vertebrates on public lands, but invertebrate and plant fossils are generally okay. When importing fossils from other countries it's best to acquaint yourself with the laws in that country to be sure you're not supporting the black market.

There are also ethical quandaries surrounding fossilized bones. If the bones are taken from their original site by non-scientists, then the chance to find out crucial information about them is lost. Paleontologists can learn a lot about the type and layer of rock the fossil was found in, its exact geographic location, and its placement in relation to other fossils and features. Most amateur and commercial fossil hunters don't bother recording such data, and once the fossil is removed that information is lost forever.

A good basic guide to fossil collecting with the above issues in mind can be found at http://www.ebay.com/gds/Buying-Fossils-Law-Ethics-Forgeries-/10000000001926697/g.html.

How to Identify Fake Hides and Bones, and Dyed Feathers

As nice as it would be to rest assured that everything on the market is the real deal, there are plenty of fake furs, bones and other items either offered as alternatives, or sold by unscrupulous sellers as genuine. Here are some tips to help you ferret out fakes:

Fur: Most brightly colored furs (pink, green, orange) are going to be fake, though there are some real furs that end up dyed day-glo colors in the garment industry. Fake fur hairs also have a tendency to clump together into triangular peaks, kind of like an anime character's hair. If you touch fake fur it often has a scratchy plastic feel to it, less smooth than real fur, and real fur often feels a bit cooler to the touch. If you can dig down through the hair to the skin, fake fur is obviously woven fabric, and you can often see lines where different colors of dye were used on multi-colored fakes. If you're absolutely stumped, try burning a few of the hairs; fake fur smells like burning plastic and shrivels up into plastic beads, real fur smells like burning hair and turns to ash. (Neither is especially pleasant, so make this a quick experiment!)

Be aware that some hides and deliberately dyed to look like others. Rabbit pelts are sometimes dyed with leopard or cheetah spots or tiger stripes; white cow hides are given zebra stripes. Most of these are labeled as such. However, if you are new to pelt identification you may need some time to learn not just how certain hides look but how they feel. A rabbit hide dyed with leopard spots is quite soft and fluffy compared to an actual leopard hide, which has sleeker and coarser hairs and feels smoother. Also, generally speaking larger animals have thicker skin, and a rabbit pelt's skin is much thinner than that of a leopard or other big cat.

Bones: There are an increasing number of good quality resin replica skulls, bones, claws, and even entire skeleton. At a glance some of them may seem indistinguishable from bone. But pick it up, and it's a different story. Resin has a smoother feel to it, even when it has texture from the skull it was cast from, and can feel heavier. And if you look closely you'll see the pores in the resin don't go as deeply as in actual bone. In some replicas the jawbone is molded separately from the rest of the skull. However, the teeth are always molded as part of the jaw or the skull; in a real skull the teeth will at least wiggle, if not come out entirely. If for whatever reason you still can't figure out what it is, use the same poke or scrape test as for the fossils.

Feathers: This isn't so much about telling a real feather from an artificial one as being able to recognize when one sort of feather is taking the place of another. Because so many feathers are illegal to have, at least in the U.S., some of those that are more in demand such as raptor feather can be had in replica form. White domestic goose feathers make a great canvas for airbrushing the patterns of hawk or owl feathers, and if you just dip the end in black ink you get a basic replacement for a golden eagle tail feather. Smaller goose or duck feathers are dyed completely black to be used instead of crow or raven feathers, though the dye makes the entire feather black; crow and raven feathers have ivory-colored quills. A few individual artists occasionally hand-paint replicas of other feathers, but there aren't widely-available replicas of, say, blue jay feathers.

So how do you tell whether you have a replica or the real deal? Generally the dye or paint lines are much smudgier than on a real feather. Also the feathers used to make the replica, such as goose or duck, have different shapes than actual raptor or corvid feathers. The book *Bird Feathers: A Guide to North American Species* by Casey McFarland and S. David Scott is a great pictorial guide to the feathers of all sorts of birds from this particular section of the world. Samples of each type of feather for each species are shown in clean detail, perfect for art reference as well as identification.

Can You Tell How an Animal Died?

Well, sometimes yes, sometimes no. If all you have is a single claw, you're likely not going to be able to tell whether it died of natural causes or not. However, if you have a hide or a skull with damage you may be able to suss out what happened depending on where the damage is and what it looks like. A hide with lots of little holes close together probably died of a shotgun wound. A large ungulate skull like a cow or horse that has a single hole in the forehead probably died at a slaughterhouse, or at least was shot on a farm as a form of euthanization. Keep in mind a lot of damage happens after death. Hides often get holes in them when being skinned without a lot of experience. Skulls can end up cracked or broken after death, especially if they're old and dried out. A lot of knowing what's the mark of the death wound is a matter of handling a lot of hides and bones and learning what to look for, both in the size and shape of the mark and its location. Many times you simply won't know how it happened unless you get the information from the person who killed the animal themselves.

There are some animal that are generally going to have died in predictable ways. Farm-raised foxes and mink, for example, are usually electrocuted or gassed with carbon monoxide. Cattle are killed with a bolt or bullet to the brain. Deer are generally shot with a gun or bow, though bone hunters sometimes find the remains of animals that died of disease and other natural causes. As always, there are exceptions to every rule, so just because you have a bone or hide from a particular animal doesn't mean that it died in the expected manner.

What if I Have Something I Simply Can't Identify?

Ask someone else. You can ask in any Vulture Culture-themed online group on Facebook or otherwise and will likely get plenty of feedback. If you have a local oddity or taxidermy shop you might try taking it in and see if they can identify it. Many universities have biology or other natural science departments whose faculty or staff might be able to help out if you email them a picture. Or try sending it to your state fish and wildlife department.

The risk with that last, of course, is that you have something you're not supposed to. If you find something outside and you suspect that it might not be legal—like a striped feather that looks like it might be from a hawk or owl—it's best to leave it where it is. Or take a picture of it where you found it and then leave it where it is until you can positively identify it. If someone does positively identify it as something you shouldn't have, put it outside. Or, if you bought it from a shop, return it and let them know that it's not a legal item to have.

Chapter 4: Here's Where It Gets Messy: Processing Raw Specimens

"Can you help me tan my first hide?"

"I found these bones, how do I clean them?"

"I have a roadkill opossum, what do I do with it?"

Sorry. Despite making art out of hides and bones for over two decades, I haven't had extensive experience with processing other than dry-preserving legal bird wings. Prepping raw specimens is messy, time-consuming, and in my Portland apartment my housemates wouldn't like the smell very much[17], while at my studio on the coast the bears would like it a little *too* much. Plus I'd rather spend my time making art with hides and bones that are already prepped.

Wait! Don't close this book! I can still help you! Just as some artists like to stretch their own canvases and mix paints from dry pigments, I know plenty of Vultures want to preserve their own animal remains. So since I'm not going to be much help with the tanning and the cleaning, I've enlisted the help of a few of my fellow Vultures, and we've written tutorials on the following processes:

- Skinning a carcass and preparing the skin
- Hair-on tanning a fur hide
- Bone cleaning
- Skeleton articulation
- Mouse taxidermy
- Wet specimens
- Dry preserving wings

Preparing a Workspace

If you're lucky, you already have a garage or other space that has plenty of room for you to make a mess. Feel free to arrange that space however works best for you, whether that's with a single workbench, or spreading everything out on the floor. (Just make sure you don't trip on anything!) It's recommended that you try and keep as much indoors as possible, especially if you are in a more rural area with bears or other large, curious critters that may investigate your goings-on. If you do some outdoor work, make sure you can shelter items from the weather or animals as much as possible.

What if you're in an apartment or only have a single room in the house? Well, a lot depends on who you live with. My partner spent years living in a small two bedroom apartment with me, and he just got used to elk hides in the bathtub and coyote skulls in the sink. Since I don't process raw items, it wasn't that messy and didn't take up that much space for very long, as I was usually just rehydrating hides or giving bones a quick bath. However, it's important that you respect the comfort of other people who use common areas like kitchens and bathrooms. Talk it over, and see what their boundaries are, and then go from there. You may just have to wait til you get your own space before you go skinning whole carcasses, but some of the projects here can be done in the comfort of your own bedroom.

Finally, make sure you aren't spilling your work out onto other people's property, or public space, without permission. A few years ago an unwitting person was startled to find several containers of macerating bones in a public park.[18] While I can understand that someone without their own space might find this to be an acceptable solution, even the best hidden caches can still be found by dogs and wildlife, and I really don't recommend subjecting other people to something they may find really disturbing. I'll talk a bit more about work space—and other space—in the next chapter.

A Few Words on Safety

As I mentioned earlier in the book, handling fresh remains in particular is risky; you should always assume that there is a chance of disease transmission and that all body fluids are similarly contagious. Be especially careful if you have any open wounds, especially on your face, hands or arms.

Some of these tutorials utilize a variety of acids, alcohols and other chemicals and generally they have their own directions included. Please make sure that you safely dispose of all chemicals used, and use only as much as you need. Rubber gloves are a must-have; the heavy rubber dish gloves that go up to your elbows are especially good, but even disposable gloves will work in a pinch. It's also a good idea to wear a respirator or other filtered mask, and goggles or other eye protection, as blood and other fluids can be sprayed remarkably long distances, and remain as tiny particles in the air longer than is healthy to think about.

How long can a carcass carry diseases? Quite some time. As long as there's wet, rotting meat on the bones there are bacteria and fungi actively breaking it down, and some of these can cause human diseases. Whitening bones with hydrogen peroxide will also disinfect the outsides, but the marrow inside may stay wet long after the outside is clean, so be careful when cutting or drilling them. You can be a little more cavalier about just carrying dry, clean bones around; beyond a certain point dangerous pathogens can't survive without something to eat. It's still a good idea to wash your hands afterwards.

Be especially careful when skinning or otherwise using sharp tools on a fresh carcass or raw hide. If you cut yourself you can be in for a really serious infection; clean any shallow cuts immediately, and deeper cuts or any punctures should be looked at by a doctor immediately. You should also have an up to date

---

[18] See: https://www.assercourant.nl/algemeen/452401/hondenbezitster-doet-lugubere-vondst-in-pittelopark-assen.html

tetanus booster, and the extra paranoid may wish to get a rabies vaccine.

Finally, make sure and clean both yourself and your working area, tools and clothing thoroughly. Scrub the work area down with a bleach solution, sterilize your tools, wash your "Vulture" clothes by themselves in hot water with a little bleach, and give yourself an extra-long shower. Some people swear by using a barrier cream (with gloves) before beginning to process specimens to help prevent pollutants from sinking into their skin. If you're having trouble getting the lingering smell of rot off your skin and hair, try using soap with lemon or orange in it, or even scrubbing with straight lemon juice. Rinsing your skin with vinegar may also help, and some people claim scrubbing with coffee grounds, toothpaste, baking soda or olive oil may be effective. And of course the quicker you can clean up, the less likely the Eternal Stench will set in.

Again, the picture quality is not what I would have liked, so color versions of all photos in this chapter may be found at http://www.vultureculture101.com.

The Basics of Skinning
By Ashley Cheeks

Skinning is an exciting process for any new vulture as it's the first major step in animal processing and cleaning. The skin is the most versatile part of the animal you can salvage as the finished product be used for a variety of projects, such as taxidermy, headdresses, clothing, crafts, and more. In addition, skinning an animal can speed up other processes. For example, the less skin there is on a carcass, the faster and more efficient bone cleaning becomes. In this essay, we will cover the basics of obtaining, evaluating, and skinning your first animal.

Where to Find Your First Project

Before you can skin an animal, you must find an animal carcass. The most immediate way to obtain a carcass is usually through salvaging roadkill. You will first have to familiarize yourself with your local laws however, as roadkill pickup is illegal in many states and provinces without a permit or is banned all together. If pickup is legal in your state and you find a carcass, you must follow basic traffic safety and hygiene. It's important to only pick up roadkill that is easily accessible (i.e carcasses on the side of the road where you can safely park your vehicle, preferably in a low-traffic area). Keep disposable gloves, plastic bags, and hand sanitizer in your vehicle.

The gloves prevent contact with parasites and bacteria on animal carcasses, and the bag puts a layer between the carcass and your vehicle to keep it clean. It should be noted that roadkill is typically more difficult to skin due to the damages animals can sustain during impact.

If you cannot safely or legally collect a roadkill carcass, there are other ways to get specimens. The best secondary option would be to order a carcass online. Once again, you must check your local laws as it may be illegal to import animal carcasses or import certain species to your area without a license. Forums such as taxidermy.net deal in whole carcass sales from hunters and trappers. Feeder animals from pet stores are also a great source for small skinning projects like mice and rats. If you are ordering online, the carcasses should be shipped as quickly as possible, no longer than a 2-3 day shipping option. Any longer and the animal may begin to spoil, making it difficult to salvage the skin. Fresh specimens are much easier to work with than roadkill.

Preparing Your Workspace

Now that you have a carcass, you must prepare a space to skin the animal. The ideal workspace is a clean, flat, easy to wipe surface, where you can comfortably skin the

animal standing or sitting. Working indoors is best, such as in a shed or garage, as insects will be attracted to the scent of the carcass and can be a nuisance during the skinning process. You'll also want to be in a space where you can easily wipe the floor, as blood and flesh may drip off your skinning station.

The tools needed for skinning animals include the following: disposable medical gloves, towels, and a scalpel with blades. Knives are a fine substitute to scalpels but are usually more difficult to use around delicate areas such as paws and tails. Other tool suggestions would be a hand flesher or a spoon for scraping excess fat and flesh after the hide has been skinned.

First Aid

Before you make the first incision you must also take your own safety into account. It's important to skin away from yourself in case your blade slips and injures you. Have band aids and disinfectants close by to deal with minor cuts and injuries. If you do happen to injure yourself, you can safely clean the wound with hydrogen peroxide to avoid potential infection. If a more serious injury does occur during the process, or a wound looks infected, seek professional medical help immediately.

How to Evaluate a Carcass

Now that you have a carcass, a work station, and your tools gathered, you must assess the state of the carcass and decide the intentions you have for the finished hide. Evaluating the carcass and skinning the animal with an end goal in mind will help a new vulture pick the right skinning method.

Here are some general considerations to take note of before skinning any animal:

Salvageable – First, you must check to see if the hide is even able to be skinned and salvaged. If the skin is extremely discolored or rotten, most of the fur has fallen off, the carcass has been sitting out in extreme weather conditions for hours or days or has tons of carrion that has been feeding on the hide, it's best to skip the skinning step all together and go straight to bone cleaning. Skinning is time sensitive, and the longer you wait to skin or the longer the animal has been left to decay, the less likely you'll be able to salvage the hide.

Incisions - The types of incisions made during the skinning process determines what purposes the hide can be used for. There are three main types of incisions that

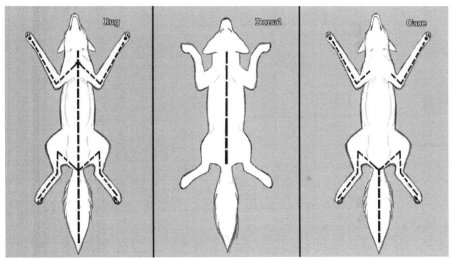

are made to skin out an animal. These include, but are not limited to, rug cuts, dorsal cuts, and case cuts.

The rug cut is for animals that will be used as flat rugs for display or for headdresses and is normally done for large animals such as deer or bear. This is because the rug cut yields the most surface area for skinning, as the incisions open the entire underside of the animal. The open incisions will allow the processor skin on large animals quickly. It's also good for roadkill that is close to being unsalvageable, as the skin can typically be removed faster than the other cuts.

The dorsal cut is a single incision down the spine of the animal and is for animals that will be used for taxidermy. This cut allows the skin to be slipped over a taxidermy form with ease and requires the least amount of sewing work once the hide has been stretched over a form. The cut can be used on any small to medium-large animals that is intended for taxidermy purposes only. It is not the best cut for display, as when a tanned dorsal cut animal is hung on a wall, the incision down the back can hang open, exposing the inner leather.

The case cut is intended for display but can also be useful for taxidermy and crafts. This cut allows the most flexibility for a new vulture, as the finished piece can be used for multiple purposes.

Taxidermy vs. Wallhanger - Taxidermy quality hides are prepared differently than wallhanger quality and how much of the animal can be salvaged is important to consider when making this decision. Animals with lots of damage to the skin, that are already partially rotting, or are just in late stages of decay are unable to be prepped for taxidermy. It's also important to note that taxidermy quality hides take longer to skin than display quality ones, and if the skin is already rotting you're on a much shorter salvage time limit.

Parasites – After an animal dies there can still be fleas, ticks, ants and other small animals on the carcass. It is recommended to freeze any specimen for at least one month prior to skinning the animal to kill off any parasites left on the carcass.

Slipping – Slip or slipping is when you tug the fur on an animal and it peels off the carcass with little to no effort. This will leave bald patches on the animal and will eventually become bare leather when the animal is tanned. Due to being exposed to the elements, roadkill is much more likely to slip, but fresh animals can as well if they were not stored properly prior to skinning.

Green Belly and Skin Rot – Green belly is rotten skin that appears green or black on the carcass, especially around the stomach. Any green or black skin will be more prone to slipping and needs to be treated with bactericides. The best bactericide on the market is called 'Stop Rot' and can be ordered from taxidermy suppliers.

Diseases – whether hunted, farmed, or fond as roadkill, not all animals were healthy before their death. Though most die with the animal, there is still the potential for certain illnesses to transmit to humans, such as rabies. These diseases can still be transmitted to humans even after freezing the carcass. Remember to always use gloves.

Other Damages – bruising, broken bones, dry skin patches, and road rash are examples of other damages found on both fresh and roadkill animals. Skinning around these damages can be difficult for beginner vultures and takes practice to get a complete hide off.

Time – this is the key difference between skinning a fresh animal and a roadkill animal. Because roadkill is often far more damaged them fresh animals, you can expect to spend up to twice as long skinning, prepping, and drying the animal.

Skinning Step by Step

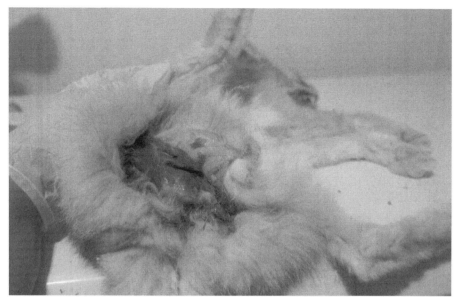

Once all your assessments have been made and you know what intentions you have for the finished hide, it's time to skin the animal. This example will be of a fresh animal that is being skinned for the purpose of taxidermy.

Start with an incision from ankle to ankle, around the anus, following the line of the fur. The next incision is down from the anus through the middle of the tail.

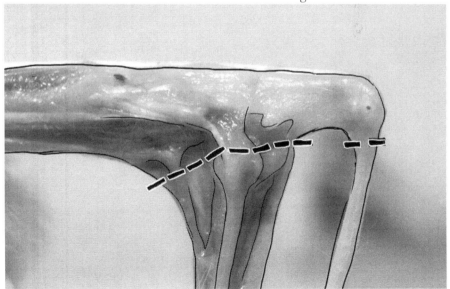

Carefully skin around both legs until the fur is loose from the ankle, then use the scalpel blade to separate the ligaments connecting the tibia and fibula to the talus bone on both hind legs. This is the space in between the bones where the leg and ankle meet.

Take the scalpel and split the paw pad, then skin around and over the toes until you reach the final toe bone. Again, use the scalpel to separate the ligaments and remove the feet from the skin.

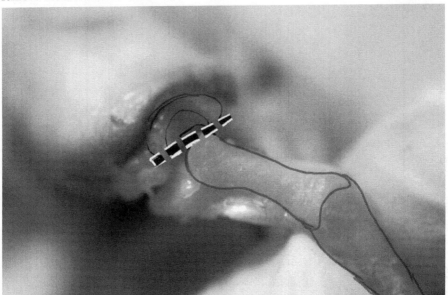

At the tail incision, cut through the ligaments in the tailbone to disconnect it from the spine.

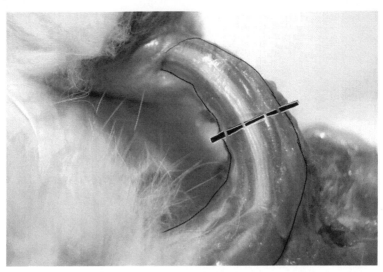

Split the tail down the middle and continue to carefully skin around the tailbone until the entire tail has been removed.

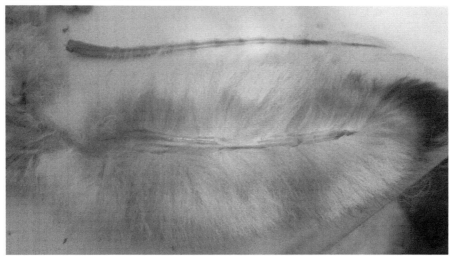

Once the feet and tail have been removed, start skinning the thighs and hind quarters of the animal. You can take your thumb where the skin meets the muscle and push the skin up toward the shoulders or bunch up the skin in your hand and pull the skin up. This process is called peeling and can be done if the skin is fresh and not damaged. Use the scalpel to cut and loosen any skin that doesn't peel after a few gentle tugs.

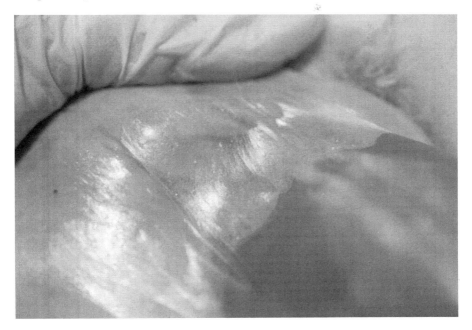

Be careful around the stomach, as most animals have very thin skin and it is very easy to cut or rip holes in the hide. It's also possible to pierce the organs in the abdominal sack, which can rupture the stomach or intestines and cause a mess. To avoid this, gently skin upwards close to the skin layer with your blade. Once you skin past the stomach and up towards the ribs, you can continue to peel and pull the skin up until you reach the shoulders.

The last incisions are from the animal's wrists down to the elbow on both front limbs. In the same fashion as the feet, skin around the forearm to loosen the skin, then use the scalpel to cut though the ligaments connecting the carpals to the radius and ulna (the area between the wrist and the arm bones.)

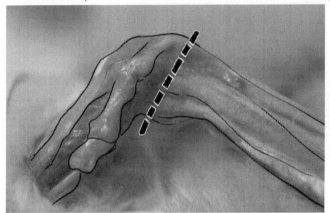

Skin up and around the wrist, separating the toes and removing the paw. Remember to pay attention when skinning around the dew claw. If you do not separate the claw from the toe in the same fashion as the other toes, the claw will not be removed with the rest of the hide and it will leave a hole in that part of the skin.

From this point you may continue skinning and pulling up the hide until you reach the cranium. Cut through the ear canals on both sides and continue working your way upwards towards the eyes.

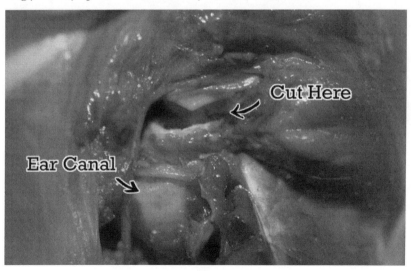

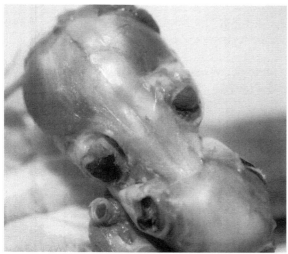

Cut as close to the eye as possible to allow the maximum amount of eyelid skin to be removed with the hide. Skinning as close to the bone as possible keeps the eyes on the pelt from enlarging, and the extra eyelid skin is necessary if the animals is to be used for taxidermy.

Now, turn the animal right side out, and use the scalpel to skin around the gum line of the animal. This line is easier to see from the exterior then the interior of the animal and makes it easier to skin the head out the rest of the way.

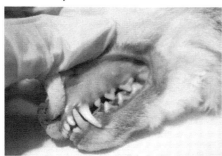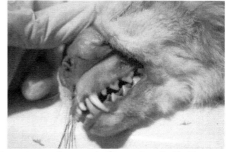

You may also separate the nose at this point to loosen it from the skull. Cut at an angle that follows the natural shape of the nasal cavity as to ensure you loosen as much material as possible. Finally, turn the animal inside out and skin out the rest of the head. You now have a complete pelt.

There are a few extra steps to prep the animal for taxidermy. The gums, nose and ears must be turned inside out to expose the excess skin, fat, and cartilage. This extra skin is needed for tucking, or pushing the skin into a taxidermy form to make the finished mount look more life-like. These steps are optional if you are solely skinning for a wallhanger, but the benefits of turning are

worth the extra time.

To turn the lips, follow the gum line incision you made earlier and skin right between the gum line and the face, gently pulling down the loose skin until you reach the line that connects the outer lip to the inner gums. The gums and base of the whiskers should now be exposed. This provides ample skin to work with when mounting an animal for taxidermy.

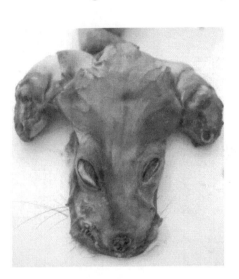 

To turn the ears, skin at the line that connects the cartilage from the skin to the back of the ear. To find the line, look for where the darker skin lines up with the lighter colored cartilage and gently skin across that line until you reach the tip of the ear. Turning ears allows for more through salt distribution during the drying process and will prevent the delicate ear skin from slipping and losing hair.

Preparing the Hide for Tanning

Once the animal has been skinned and prepped, the hide must be scraped to remove large portions of meat and fat still present on the skin. This process is called fleshing. Traditionally, fleshing is done over a long wooden board called a fleshing beam and scraped down with a fleshing knife. For the beginner vulture, you can use a dull scalpel or even a spoon to flesh a good portion of the meat and fat off smaller animals.

Finally, after the hide has been thoroughly fleshed, it is now ready to be salted and dried. Turn the animal fur side in and rub salt all over the skin. Make sure to salt every inch of the animal, as any unsalted skin has the potential to slip and rot. If able, it's best to hang the salted skin in front of a fan to dry it faster. Within a week the salted skin should turn from the raw skin color to a white color and be very stiff. The skin is now complete and ready to be tanned.

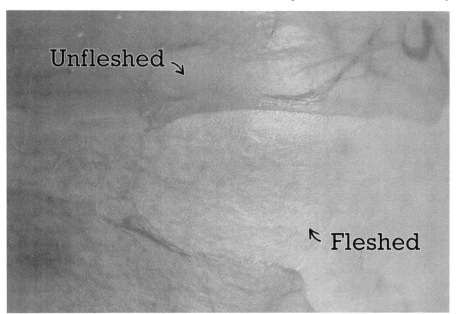

For a new vulture, the task of skinning can feel overwhelming. However, like any other type of animal processing, it becomes much more fluid with practice. Work slowly and carefully, paying attention to each step. With patience, dedication, and a steady hand, any vulture can become an expert and skinning and preparing animals.

Hair-On Tanning
By Eric Foote

Tanning is the process of altering the structure of an animal skin in order to permanently preserve it. Skin is composed of two types of protein: fibrous proteins, which form the structure of the skin, and globular proteins, which are non-structural. The main structural protein is collagen, and that is what the tanning agents bind to. The globular proteins are un-tannable; they are simply in the way, taking up the space in the collagen structure that the tanning agents must bind to. The majority of the tanning process is composed of prep work on the hide to remove these globular proteins, which will then allow the tanning agent to bind successfully to the collagen structure. This bond changes the collagen from a protein state to a non-protein state; it becomes a stable material (leather) that is more durable and will no longer decay in normal environmental conditions.

There are many different methods of tanning, but the best and most beginner-friendly is modern synthetic tanning. The chemicals are specifically formulated for their purpose and will therefore produce the highest quality, most consistent results. This tutorial will use Rittel's brand tanning products, which can be purchased online at TaxidermyArts.com.

A lot of time and effort is required in order to properly tan a hide! Expect the process to take one to three weeks or more from start to finish, depending on the size of the animal and how much time you're able to devote to working on it each day. One major advantage of this method is that there are several points at which the hide is preserved in a stable state and can be left for extended periods of time, allowing you to work at whatever pace is best for you.

Necessary Tools and Chemicals:

- Disposable gloves
- Scalpel with replacement blades, or a knife that can be sharpened
- Two plastic buckets or tubs with lids, large enough to contain the hide you are tanning
- Kitchen scale
- Measuring spoons (teaspoon and tablespoon)
- Measuring cup (one quart/four cup size or larger is recommended)
- pH test strips (low range, from 0.0 to 6.0, is recommended for more accurate readings. These are most easily found at online retailers such as Amazon)
- Salt (can be found in bulk at Tractor Supply Co., most livestock feed stores, and bulk retailers such as Costco and Sam's Club. Get fine-grained salt, not rock salt)
- Sodium bicarbonate (baking soda)
- Rittel's Saftee Acid
- Rittel's Super Solvent Degreaser
- Rittel's EZ-100 Tan
- Rittel's ProPlus Oil

Optional, But Recommended, Tools and Chemicals:

- Fleshing beam (can be made, or purchased from a trapping supplier such as FNTPost.com)
- Fleshing knife (can be found at trapping suppliers such as FNTPost.com)
- Sandpaper (100-150 grit)
- Rounded sanding block
- Rittel's Ultra-Soft Surfactant

Step 1: Fleshing

Fleshing is the removal of the fat and meat from the skin. This is most easily done on a fleshing beam with a fleshing knife, but can also be done with a scalpel or sharp knife if you don't have access to those tools.

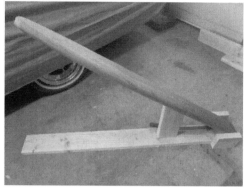

A fleshing beam can be fashioned from a sturdy wooden board sanded to create a rounded surface, a round wooden post sanded smooth, or a PVC pipe. A good size that works for most animals is five to six inches wide by fifty to sixty inches long. It can either be used by laying it on a table and bracing it between your body and a wall, or building a base to hold it up at an angle that is comfortable for your height.

If you're using a fleshing knife and beam, make sure to brush out any burrs that may be stuck in the fur, as they can cause the knife to snag through the skin and cut a hole as you scrape over them. Slip the hide over the beam, hold it in place by pressing your body against the tip of the beam where the head of the animal is, and begin scraping behind the ears toward the shoulders. Hold the blade nearly flat against the skin, at only a slight downward angle. Work your way down the entire hide, rotating it as you go to get all sides. The flesh should scrape off pretty easily. For any tougher areas on thicker skinned animals, like around the neck and shoulders, it can help to first start fleshing it down a little ways with a

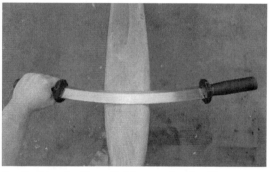

scalpel/knife, then going back to the fleshing knife to finish scraping it off.

If you're using a scalpel/knife, lay the hide out on a table, hold the blade nearly parallel to the skin, and carefully slice the meat and fat away, cutting in between the flesh and skin. Keep the skin taut using your other hand, so that it doesn't bunch up or wrinkle and cause you to make holes. Fat can also easily be scraped away with something rounded and dull, like a spoon.

You don't need to worry about removing every single tiny bit of flesh and membrane at this point, just make sure to remove any chunks or large sheets of flesh.

The following pictures show a raccoon hide before and after fleshng.

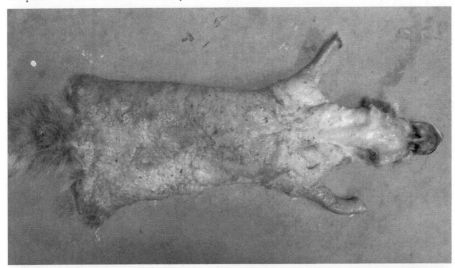

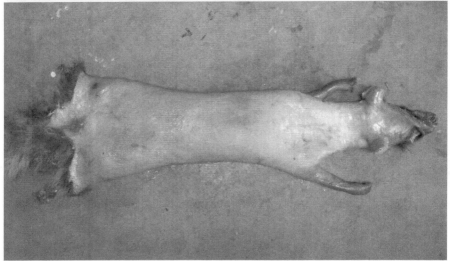

Step 2: Salting

You will need finely granulated plain salt, just like the table salt used for seasoning. Both iodized and non-iodized are safe to use. Do not use coarse or rock salt. The larger grain size of coarse salt doesn't allow it to cover the skin or absorb fluid as effectively. Rock salt has the same issue, as well as having more impurities that may react with the chemicals used later on. Do not use anything other than salt. Borax is another product sometimes recommended for drying out raw hides, however it should not be used on any hide that is intended to be tanned. Borax has a pH of 9.3, which is unsafe for the skin at this stage and can cause problems during the tanning process. When a hide become too alkaline the collagen degrades, loosening the hair follicles and causing the fur to fall out. The chemicals used in the tanning process are also pH-sensitive and any borax left on the skin may alter the pH of those solutions and reduce their effectiveness.

Lay the fleshed hide out on a flat surface, and rub the salt onto every square inch of the hide, making sure it gets into every crease and fold. Keep adding salt until you have a layer just thick enough that you can't see the skin underneath. What the salt will do is pull out fluid and oils from the skin, dehydrating it to aid in preservation, as well as drawing out some of the un-tannable proteins. This will allow the subsequent pickling solution, and later the tanning agent, to work more effectively.

After twelve to twenty-four hours, scrape off the dirty salt and rub on a thin layer of fresh salt. Hang the hide up so that there's airflow on all sides. It can be hung by draping it over two horizontal poles held up by two chairs, tables, etc. on either side. Leave it to dry like this for two to five days, depending on the thickness of the skin. Allowing the hide to salt dry tightens the hair follicles, helping to prevent fur slipping (fur falling out due to bacterial activity/decomposition). Once the hide is mostly stiff and has developed a salty "crust", it's ready to move onto the next step.

After drying, the hide can be stored if you can't continue working on it right away. As long as it's kept somewhere cool, dry, and where pests can't get to it, it can be left salt dried for several months. Species with a lot of fat, such as raccoon, opossum, groundhog, skunk, etc., shouldn't be left dried for any longer than that as they might develop grease burn, which is when the fat in the skin oxidizes into fatty acids that break down the skin, leaving it brittle and unable to be tanned. Grease burned hides will darken in color, going from white or yellow to orange, and eventually to black. If you notice your hide has changed color, move on to the next step as soon as possible.

Step 3: Rehydration

If your hide was recently salt dried, you can simply rehydrate it in plain, cold water. Let it soak just long enough for it to fully relax and become soft, just as it was after skinning. Depending on the thickness of the skin and how long it's been dried, this can take anywhere from ten minutes to several hours. Keep a close eye on it and don't leave it in the water for any longer than necessary, as bacteria thrives in water and may begin to spoil the hide, causing the fur to slip, if it's left too long.

Hides that are thicker and dried hard will rehydrate faster with a surfactant added to the rehydration bath, reducing the chance of spoilage. For every gallon of warm water, add one ounce of Ultra-Soft. Allow the bath to cool before placing the hide in.

After the hide is fully rehydrated, give it a quick rinse in cold water to wash away any dirt or blood in the fur, then squeeze or gently wring out the excess water.

Step 4: Pickling

A pickling solution is a mixture of water, salt, and acid that the hide is soaked in.

The acidity kills bacteria on the hide to preserve it, as well as further breaking down the remaining non-structural proteins. It also swells and toughens the skin, making it easier to flesh and shave in the following step. Pickling is incredibly important for achieving soft, high quality leather.

To make a pickling solution, mix one half ounce of Saftee acid and one pound of salt per gallon of water. Stir the solution thoroughly to ensure all salt is fully dissolved. There should be enough solution that the hide can be fully submerged without being scrunched or folded up tightly, that way the pickle can access all areas of the skin. A good rule a thumb for the minimum volume of any solution the skin will be soaking in is one half gallon (two quarts) of solution per one pound of the skin's weight.

Before adding the hide, check the pH of the solution with pH test strips. It needs to be between 1.0 and 2.5. During the first twenty-four hours after adding the hides to the pickling solution, re-check the pH every few hours and stir the solution to ensure the pH is stable and the hide is getting pickled evenly. To maintain the pickle after that, leave a lid on the tub to prevent evaporation, and once a day check the pH and stir the solution. As long as the pH stays under 2.5, the hides are preserved and can safely stay in the pickle indefinitely. If the pH ever rises above that, remove the hide and add small amounts of acid to the pickle until it's back down to the right level before putting the hide back in. These two pictures show a fox hide before and after pickling.

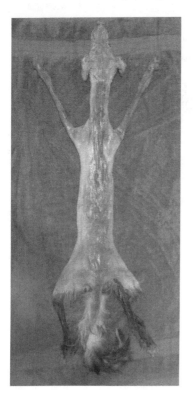 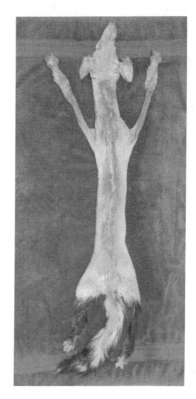

Step 5: Detail fleshing

After pickling for at least three days, the hide is ready to be detail fleshed. This means removing the bits of meat or fat that were left behind from the initial fleshing, as well as all membranes. Detail fleshing can be done with a scalpel or sharp knife. Hold the blade nearly parallel to the skin, slicing in between where the membrane meets the skin. Roll the membrane down with your other hand as you slice. Sometimes the membranes can also be easily peeled off by pushing your fingers between the skin and membrane and carefully working it down.

Take your time with this step; the skin is preserved by the pickle so you can work on it for as long as you need to without worrying about it spoiling. If you need a break, simply return the hide to the pickle. The entire skin should be smooth, with as little remaining membrane as possible.

This picture shows pickled skin in the process of detail fleshing. The left side has had membranes removed, while the right side still has membranes present.

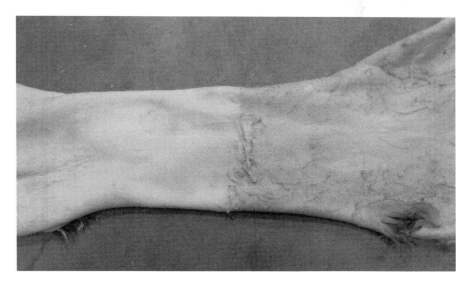

Step 6: Degreasing

All of the fat on the surface of the skin has been removed, but there are still fat and grease within the skin that need to be cleaned out. It's incredibly important that the grease is 100% removed, otherwise it will prevent the tanning agents from properly bonding to the skin and will degrade the skin over time.

If you have a very greasy animal, you will need to manually remove as much grease as possible before putting it into a degreasing bath in order for the bath to work properly. After all membranes are removed, scrape the skin with something dull like a spoon. As you scrape you will see a buildup of fat and grease on the spoon. Be sure to scrape the entire skin, even the tail, which can sometimes have a lot more fat in it than you'd expect. Once there is no longer any fat building up on your scraping tool, it can be put into a degreasing bath.

To make a degreasing bath, mix four ounces of Super Solvent and four ounces of salt per gallon of warm water.

As the hide soaks, periodically agitate it and scrub it with your hands (while wearing gloves). This will help the degreaser pull out the grease. Critters with little fat will only need about an hour in the bath, while fatty ones will need several hours. Don't leave a hide in for longer than ten hours. The acidity of the pickled skin will lower the pH of the degrease bath to about 3.0 or 4.0, enough to keep bacteria at bay for a while, but not forever. If left in the degreaser too long, bacterial activity may begin and cause the fur to slip.

After removing the hide from the bath, give it a quick rinse, squeeze out the excess water, and return it to the pickle.

Step 7: Neutralizing

Once the hide has pickled at least overnight after it's been degreased, it's fully prepared for tanning. But before it can go into the tanning solution, the high acidity from the pickle must be partially neutralized.

To make a neutralization bath, mix one tablespoon of baking soda (sodium bicarbonate) per gallon of cold water.

Remove the hide from the pickle, squeeze out as much pickling solution from it as you can, place it the neutralization bath, and stir it occasionally as it soaks. Thin skinned animals such as fox, rabbit, and opossum need only five to ten minutes, while thicker skinned animals such as raccoon and deer need fifteen to twenty minutes. Do not neutralize for any longer than twenty minutes, as the hide needs to still be slightly acidic so that the tanning agent will properly bond to it.

After neutralizing, give the hide a quick rinse and squeeze out excess water.

The pickling solution can be saved and reused for the next hide you tan, or disposed of. To safely dispose of it, neutralize it by stirring in one tablespoon of baking soda per gallon of solution before pouring it down a drain.

Step 8: Tanning

To make the tanning solution, mix one ounce of EZ-100 and six ounces of salt per gallon of warm water. The EZ-100 powder can be difficult to dissolve; it helps to first mix it in a smaller container of water so that it's easier to stir before pouring it into the larger tub.

Before placing the hide in the solution, check the pH. It should be 4.0. If it's too high, stir very small amounts of Saftee acid (about 5 drops at a time) until it's at 4.0. If it's too low, dissolve one teaspoon of baking soda into one cup of water and add that mixture into the solution one tablespoon at a time until it's at 4.0.

When placing the hide in the tanning bath, make sure to fold it into the solution so that the skin is fully submerged and isn't twisted or scrunched up tightly. This will prevent air pockets and tight folds, so that the solution can flow evenly over all areas of the skin. After only a couple of minutes of soaking in the solution the skin will change in texture, from slick and a bit slimy to coarse, almost fabric-like.

Thirty minutes to an hour after placing the hide in the solution, check the pH again. If it's higher or lower than 4.0, remove the hide and adjust the solution with acid or baking soda as necessary, and return the hide to it. If any adjustments are made, re-check the pH after another thirty minutes to an hour. If it has stayed at 4.0, it is stable and the hide can be left alone to finish tanning.

The hide is fully tanned after sixteen hours. It can stay in the solution for up to twenty-four hours, but after that the skin will begin to lose elasticity and won't stretch as well during the breaking process. Before removing the hide from the solution, the pH must be raised slightly to eliminate the possibility of acid rot (acidity damaging the leather over time and leaving it brittle). Using the same method of raising the pH described above, bring it up to 5.0 or 5.5. Leave it for thirty minutes, and check the pH again. If it has stayed at 5.0 or 5.5, the hide is ready to come out of the solution. If not, bring it back up to 5.0-5.5 and leave for another thirty minutes.

After removing the hide from of the tanning bath, rinse it well, then wash the fur with shampoo or dish soap to clean any remaining dirt and leave it smelling nice.

Step 9: Oiling

After washing the hide, squeeze and shake out as much water as possible, and towel dry the fur. Turn it skin side out and roll it tightly in between two towels, letting it sit for about ten minutes. This will wick out excess moisture from the skin, allowing the oil to absorb more effectively.

Dilute the ProPlus oil, one part oil to two parts hot water, and apply it to the skin with a paintbrush or gloved hand. Make sure to coat the entire skin, but be careful around the edges and around holes, so that you don't get oil onto the fur. If you have any excess oil mixture left over, you can seal it in a container to use on your next hide, just be sure to warm it up before use.

Once the skin is oiled, turn it fur side out, fold it up, and let it sit for four hours to allow the oil to fully absorb into the skin before it begins to dry.

Step 10: Breaking the Leather

Once the oil has fully soaked in, the hide can now be hung to dry. Be sure to dry the fur before turning the hide skin side out to dry the leather, either by blow-drying on a warm or cool (not hot!) setting, or letting it air dry. Back-brushing the fur (brushing it the opposite way it naturally lays) as it dries will keep it fluffy and soft.

What makes a hide soft and pliable is the ability for the fibers of the leather to slide across one another (think of leather as a woven fabric). The oil lubricates the fibers so they can slide more easily. As the skin dries, the fibers will stick together and make the skin stiff. "Breaking" the leather is the act of stretching and forcing the fibers apart before they become stuck, so that they remain mobile.

As the leather dries, it will turn a darker color and start becoming a bit stiff or crinkly. When you stretch it with your hands, it will turn bright white and become pliable again. Continuously check the hide every thirty minutes or so, stretch it in all directions, and generally work it with your hands (crumpling it like you would a piece of paper, wringing it, etc.) until it's completely dry. What also helps quite a bit with breaking hides is "staking" them, which means dragging the leather back and forth across a sharp edge, such as head of an axe or shovel, edge of a table/shelf, back of a chair, etc. If you have a fleshing beam and knife, you can lay the hide on the beam and scrape it with the knife to stake it as well (after cleaning the knife and beam very well, of course!). Begin staking once the leather is dry to the touch but not yet 100% dry. These photos show dry leather before and after being stretched/broken.

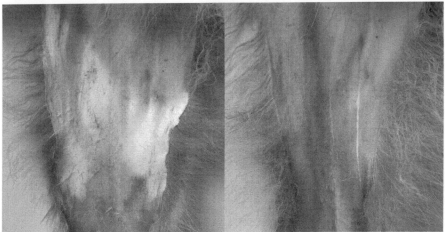

When you won't be able to work on the hide, such as at night when you're sleeping or while you're at work, it can be stored safely in a refrigerator. If it needs to be stored for longer than twelve hours, keep it in a freezer to prevent the growth of mold or mildew.

If any part of the hide dries too much before you get to breaking it, or it doesn't come out as soft as you'd like, simply re-wet the area with a sponge or spray bottle and re-break it.

A small or thin skinned animal can usually be fully dried and broken in one full day, or about twelve to sixteen hours of total drying time, depending on humidity. Larger or thicker skinned animals may need up to thirty-six hours or more of drying time.

Step 11: Finishing

The final finishing touch on the leather is sanding it down. This step is optional, but it will leave your leather with a smooth, clean looking surface as well as help further soften it.

Use medium (100-150) grit sandpaper, and sand by hand with a rounded sanding block. Lay the skin over a table, holding it in place by leaning your body against the table, pull it taut with one hand, and sand with the other hand. The more thoroughly you flesh the hide before tanning, the easier sanding will be. The following two pictures show close-ups of leather before and after sanding.

If after the fur dries you find that some softening oil did get onto the fur, it can be easily cleaned up using a small amount of the Super Solvent degreaser mixed in water. Stir one teaspoon of degreaser into one quart of water, wet a rag or paper towel with the solution, and wipe down the oily fur. You can also pour the solution into a clean spray bottle and spray the fur. Wipe away the degreaser solution with a wet rag or paper towel, and dry the fur.

You're Done!

Finally, after many hours of work, you have beautiful tanned pelt! To care for it, simply brush the fur occasionally to remove dust and prevent matting. Long term exposure to heat and moisture will degrade the leather, so keep it in a room that isn't excessively hot or humid and out of direct sunlight. If the pelt does spend some time in a poor environment that dries out the leather, use the ProPlus oil to restore it. Do not use leather conditioners such as Neatsfoot or mink oil. Leather conditioners are intended for the grain side (outside layer) of the leather which isn't as absorbent. They are formulated to create a protective coating on leather goods, rather than fully absorbing into the fibers of the leather to make it pliable like tanning oils do. A pelt that is properly tanned and cared for using this process will last for generations.

Bone Cleaning
By Shelby Hendershot

Bones: The skeletal structures found in all vertebrates. These supportive, varied and often seemingly engineered structures found in birds, mammals, reptiles, and other higher orders range from the tiny to the utterly massive, thick structures to delicate ones. Perhaps the true meaning of "the beauty within," skulls and skeletons are collected throughout the world by private individuals, institutions, and traditional cultures alike. While the reasons behind the creation of these collections are often educational, religious, or just for fun, they all have a single thing in common that makes them possible; bone cleaning. Bone cleaning takes many forms, often related to the specimen, its use, and end product. Here will we will discuss the good, the bad and the downright smelly side of current and historical bone cleaning techniques.

Specimens can be sourced from many places. Common North American furbearers, like coyotes, raccoons and foxes, are good first projects for beginner skull cleaners. They are of a nice size, not too large, not too small, readily available and inexpensive. Often larger domestic skulls like goats, sheep and cattle can also be sourced easily. Horned and antlered skulls will require some specific handling and are a bit more challenging; I'll talk more about that later. Raw skulls can be purchased from trappers, hunters, farmers, butchers and exotic animal breeders, and several companies online offer raw skulls.

Be sure to be aware of your state laws regarding raw specimens. Of special note is the restriction of the movement of raw cervids (deer, elk, moose, etc.) across many state lines. This is to reduce the risk of spreading Chronic Wasting Disease (CWD), a devastating brain disease caused by a prion similar to Creutzfeldt Jakob Disease (Mad Cow). Regulations regarding raw skulls can often be found on the website of your State's Fish and Game Department, or by emailing them directly. You can also find your state's information at http://www.thegreenwolf.com/animal-parts-laws.

Roadkill is also a good source for specimens if it is legal to collect roadkill in your area. Be sure to respect private property and be safe when collecting from public roads. Roadkill laws vary state to state so be sure to familiarize yourself with what is legal to collect in your area. Local farmers can be a good source for domestic species. Networking with livestock producers at farmer's markets, animal expos and county fairs is a great place to start. Any specimen should be stored well wrapped in a good quality plastic bag and frozen until you are ready to work on it.

No matter how your specimen was sourced the techniques and equipment to prepare your specimen for cleaning vary little. You will need:

- Nitrile or latex gloves - Worn when working with any raw tissues or chemicals.
- Personal protective equipment (PPE) - Apron, mask, face shield may also be desired.
- Watertight container - Recycled plastic containers, buckets, etc.

- A sharp knife - Knives that accept replaceable scalpel blades are my personal favorite. They offer the precision and sharp edge to get the job done, but any small knife with a good edge will work great. Exacto knives should be avoided. Their carbon steel blades are not designed to cut flesh and will dull quickly.
- Dawn dish soap - Clear, liquid. Do not use colored Dawn as it can stain the bone.
- Peroxide - 3% hydrogen peroxide, found in the First Aid section of any grocery or big box store.
- Paper towels - Heavy duty, free of any printed designs.
- Plastic bag - Trash bags or zipper seal food bags are preferred.
- Fine screen - Plastic or fiberglass, fine mesh on a frame. Old window screen works well.

Optional Tools:

- Brain removal tool (specifics will be discussed in a moment)
- Heat source

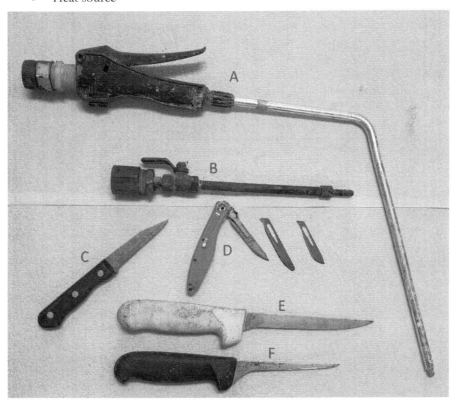

Preparing Your Specimen

First a specimen should have all skin, hair, and feathers removed, with very few exceptions that will be discussed later on. If the skin of the face is not to be kept for tanning, it is often easiest to make an incision from the center of the neck the end of the lower jaw and skin up either side of the skull. Ear cartilage should be cut

off as close to the skull as possible. Care should be taken when skinning to not score or cut the bone with your knife as these cuts will show up on your finished specimen. Cut with your knife blade parallel to the bone and separate the connective tissue between the skin and bone. When skinning horned game, be sure to make a good clean cut around the base of each horn just above the hair/horn line. The horn tissue in this area is very soft and easily cut. A smooth cut here will save you having to touch up your horn edge later on in the process. Antlered game also requires some finesse around the base of the antlers, called the pedicle. Make a cut on one side of the pedicle or horn and using the tip of your knife carefully cut the skin off the bone, making sure to get as much hair as possible. Skulls of smaller animals, like mink and muskrats, rarely need additional prep work beyond skinning.

After skinning larger specimens, it is a good idea to cut off as much of the soft tissue from the skulls of anything raccoon-sized or larger. Remove the eyes, tongue, and large muscles on either side of the skull and jaw. Skeletal sections should be treated similarly removing as much of the bowel, muscle, and tendon as possible. Treat hooves of small and large ungulates just like a horn.

Depending on your preferred method of cleaning you may wish to remove the brain at this time. Brain removal is NOT required, however it can speed up the cleaning process, and since brain tissue is very fatty it can cause problems later in the process, especially when using beetles. It can be easily removed without damaging the skull using water pressure. A simple tool can be built with parts from your local hardware store. You will need a valve that is adapted to fit on a garden hose, a reducing fitting, and a four to twelve-inch long tube that is ¼" in diameter. If you plan to do nothing larger than a deer an eight-inch tube will suffice. Your valve, reducing fitting and tube should connect to make a tool that resembles an oversized garden nozzle. Brass or copper are best, however, rigid plastic can be used for light duty. The tube of this brain removal tool is carefully inserted into the foramen magnum, until it touches bone. Carefully turn the valve to force water into the skull, behind the brain. Hydraulic pressure will then push out the softer brain tissue. Wear an apron for this step. A face shield is also recommended until you get the hang of it! After removal of the brain we recommend rinsing off the skull, including rinsing out the nasals to remove any food, bugs, sand or other debris before cleaning.

North America has several species of horned game, including the American Bison, Mountain Goat and several species of wild sheep. Many domestic animals can also sport these often-impressive head adornments. Horns are beautiful and add extra majesty to a finished skull; however, they do require extra attention during the cleaning process. The horn is made up of three main parts. The outer keratin sheath which is similar to claws, fingernails and hooves. The second layer of a horn is a thin membrane of soft tissue full of blood vessels that joins the sheath to the bone below. The bony innermost layer called the core is the extension of the sinus cavity and is quite hollow. If a horn sheath is not removed from its core, the soft tissues will quickly begin to smell, putrefy and attract insects. Hooves and claws are similar and should be treated just like horns during the prep stage.

The most effective method for removing horns is to sweat them off. After skinning and rough defleshing, wrap the entire skull, and if possible the horns, in a trash bag and place in a warm location. If the horns cannot be bagged, use duct tape at the base of the horns to seal the bag. Be sure the tape does not touch the horn itself as the adhesive can be very hard to remove. After a few days pull on each horn and wiggle them. If they do not move, rewrap, and allow it to sit for a few more days. Once enough decay has happened the horn sheath will pop right off. The sheath can then be washed in hot soapy water and filled with peroxide for a few hours to kill bacteria. Horns should never be partially or fully submerged in peroxide as it can quickly cause discoloration to the outside of the horn. After washing and sanitizing, horns can be allowed to dry at room temperature, out of direct sunlight. Store horns in a dry place until your skull is finished. After horn removal but before cleaning, if desired, you can trim the bony cores to one half or one quarter their length to help fit the specimen into a smaller container. Be sure to leave enough core to support and properly align the sheaths after cleaning.

After prep and horn removal specimens can be frozen and stored indefinitely until you are ready to clean. Be sure to wrap and pad the teeth to avoid damage during storage, four or five layers of paper towels is sufficient.

Maceration

The technique most readily available to beginning Vultures is maceration. Maceration is the process by which water and bacteria in suitable conditions team up to break down all soft tissues on a specimen. This process, when properly managed, does not damage bone, preserves every tiny detail and is very inexpensive. This technique requires, at its minimum, little more than a container that holds liquid, a specimen, and tap water.

The first step for successful maceration is to choose a safe location for your work. A secure fenced in area away from wildlife, domestic animals and pets is best. Secure fences can be built with plywood, stock panels, woven wire, etc. If you live in a high traffic area for coyotes, badgers, or similar scavengers you may wish to add a "floor" to your area to prevent unwanted animals digging in. Best options for a floor is something strong like plywood or chain link, securely attached to your fence. As long as your tubs are covered, birds will not be an issue. This netting, typically used to cover bird pens and fruit trees, is available at most hardware stores. Enclosed spaces like a garden shed also work well, but only if you plan to use an artificial heat source. Passive maceration, described below, will be less successful in a shed or garage. These dark buildings will not warm daily with solar energy, like a tub in direct sunlight would. Greenhouses are ideal locations for passive maceration. They collect the solar energy and keep the ambient temperature warmer, longer.

Choose a container that is large enough to allow you to completely cover your specimen in water and the ability to swirl the bones or skull in the container. Containers with lids are best; recycled peanut butter jars, deli food containers and durable ice cream tubs are my personal favorites. Larger specimens can be contained in one to five-gallon buckets, plastic totes, and livestock water troughs. Metal containers or containers with metal lids should not be used at any point in the process. The metal will rust and can discolor or cause damage to a specimen later in the process. Glass can be used but it is not preferred for safety reasons, as it is easily broken if dropped. A good sealing lid is helpful and will reduce evaporation and the risk of spills. In lieu of a lid you can use a layer of plastic bag floating on the surface to reduce evaporation.

Add your thawed specimen to your container and cover it in warm water. Using cold water is okay but won't really get you a jump start on the process. Do not use hot water as sudden temperature changes can cause teeth to crack.

Maceration is managed two ways, described herein as active and passive. Active maceration is heated. An outside heat source warms the water to between 85 and 105 degrees which encourages rapid bacterial growth and efficient cleaning. Heat sources can include (but are not limited to) aquarium heaters, heated mats, light bulbs, and water heating elements. Always check the cord and element for damage before using. Only plug heaters and appliances into approved GFI (Ground Fault Interrupt) outlets when working around water. Passive maceration relies on ambient temperature of the air around it to encourage bacterial growth. Typically done at room temperature or outside temperatures, this process can be improved by placing the container in a location that receives a lot of sun or a warm location in a home. Outside locations like greenhouses or southern exposures are great. Inside the decay process can be sped up by placing well sealed containers near your home's heat source. Remarkably, a container with a good lid gives off little to no smell. Central heat floor vents are great, as are hearths if the container is placed at a safe distance. Passive maceration will speed up and slow down as the ambient temperature changes. At air temperatures below 50 degrees Fahrenheit the process slows to a crawl. It is best to prevent any water submerged skull from freezing. Water swells as it freezes and this can really cause significant damage to bones and skulls.

The process of active maceration can take a few days or weeks, compared to several months or years typical of passive maceration. Time varies with the size, species and health of the specimen. Healthy animals tend to hold more fats and oils in their soft tissues and bone marrow. Comparatively, a specimen that succumbed to a long illness has likely consumed its fat reserves, making for much less grease in the bone. Long cleaning times with passive maceration can cause some degeneration in bone so it is best to check specimens for damage regularly. This damage can present as splits, cracks, or algae growth. Long term soaking will also cause separation in the sutures of a skull. These natural seams in a skull allow for growth and will not be fused in younger specimens. Consistent warm temperatures are key to a quality maceration process. A digital thermostat with a remote thermometer is helpful for managing and maintaining a productive maceration.

After two or three days with active maceration you should swirl the specimen in its container and check progress. If it is not visible through the fluid or you notice large amounts of dissolved tissues you can carefully drain off the fluid to check progress. Don't rinse it with clean water, just drain off the old fluid at this point and refill. You want to keep some of the gunk to help quickly reestablish a bacterial population. Fluid should be poured through a fine screen (relative to the specimen you are working with) to ensure no teeth or small parts are lost. If a specimen is slow to clean, allow more time before checking progress again.

Once a specimen appears to be free of soft tissue after draining fluid, carefully check the teeth and remove any loose ones. This helps the bacteria clean the tooth sockets and interior of the jaw and upper skull and helps with the degreasing process later. Not all teeth can be removed: this is okay. Be sure not to lose any teeth during water changes and rinses. Thoroughly rinse the skull including swishing out the brain case, refill the container with warm water and allow the skull to soak for several more days. Once the water is clear-ish after several changes and no longer super smelly a specimen is ready to proceed to the next step. This process of water changes applies with passive maceration as well. Just allow longer times between checks.

Be aware of evaporation in open top containers, refill the water as needed. Sometimes during cleaning specimens will change color, most commonly to a strong black color. Don't worry, this color will be removed later in the process. If a specimen begins to grow algae (green or red) this means you do not have an active bacterial growth. You need to move your specimen to a warmer location and encourage bacterial growth. Covering your container with a black trash bag or dark lid will help prevent algae growth and heat the liquid. Liquid slurry from an active maceration or well

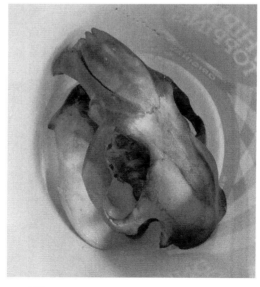

performing passive maceration can be added to an underperforming maceration to speed things up. Some recommend adding meat to a maceration to speed up the process, this works, but is not as fast as adding the liquid slurry. The slurry disperses much quicker throughout the specimen.

During cleaning you may notice a white crumbly substance developing on your specimen, especially on very fatty species like bears, pigs, and domestics. This is fat that has "saponified" and turned to a substance called adipocere or coffin wax. This generally means that your maceration is too cold. Large amounts of adipocere should be removed as they form, best done with a stiff brush or dental pick. Small amounts of adipocere can be removed later in the process. If you are seeing large amounts of adipocere formation regularly with active maceration it can mean that you are not removing enough of the soft tissues during the prep stage. Maceration water does not contain any chemicals. Never add soap, peroxide, borax or any other thing to a maceration. Recently it has come up in bone cleaning circles that septic tank enzymes can improve maceration. These products are not recommended for bone cleaning as the ingredients vary by brand and they can cause damage or slow the natural bacteria growth impeding the process.

Maceration water can be poured down the drain or toilet. You can also use this nutrient rich liquid to improve compost or soils. If poured outside it will have a heavy smell until it breaks down so be conscious of neighbors. One consideration for maceration water disposal is with specimens euthanized with phenobarbital, a strong chemical that does not break down. Water from euthanized specimens should not be used for soil improvement; it should only be poured down the drain into an approved septic tank or city sewer where it will be handled in a waste treatment facility. Incineration is the preferred disposal method of any leftover remains of euthanized animals, but most landfills will accept them as well. Euthanized animal remains should never be cleaned by a method or disposed of in a way that allows scavengers or pets to reach them.

After maceration is complete, specimens should be scrubbed in hot, soapy water. Clear Dawn dish soap is my preference; any dyes, especially blue, can stain bone and these stains are nearly impossible to remove. If your specimen is large or from a mature animal don't be afraid to put some elbow grease behind scrubbing, especially if it has coffin wax. Be sure to rinse it well. Wash out your container, or switch to a new one, and refill with water for storage until the next step. Storing specimens in water prevents mold growth and stains. Be sure your specimen is covered, and all the teeth are included. At this point you can either go straight to whitening or proceed to degreasing. Younger specimens, especially very young juveniles and fetals, rarely need degreasing. Older, mature animals usually require some degreasing. Thicker, denser bones like the long bones, jaw and rear portion of the skull are more likely to harbor grease and will require more time.

To degrease bone, use one-part clear Dawn dish soap to ten parts hot water. Mix well and fill the container to cover the specimen. When degreasing with dish soap, it is important to keep the water warm, 80 to 110 degrees Fahrenheit is ideal. Heat the water the same way you did during maceration. Degreasing takes from a week or more for small bones to several months for larger, greasier ones.

Water should be changed when it becomes cloudy or scummy. This may be daily or weekly. Once the water stays clear for a week or more, your specimen is ready to proceed to whitening.

More experienced Vultures may use acetone for degreasing. Acetone is extremely flammable and should never be heated or kept near a heat source; it also requires special disposal and should never be poured down the drain or onto soils. Always familiarize yourself with the MSDS (Material Safety Data Sheet) of any chemical you plan to work with; these are easily searched online.

After your specimen is fully degreased, proceed to whitening, a process by which bone is sanitized and lightened in color. This process also helps remove any remaining foul odors and bits in the dark recesses of your skulls and bones. Only one chemical is used to safely whiten bone, hydrogen peroxide, $H2O2$; 3% hydrogen peroxide can be easily obtained in the First Aid section at any pharmacy or grocery store. It is typically packaged in one-quart brown bottles. This concentration is sufficient for nearly all bone whitening. Enough peroxide should be purchased to completely cover your specimen in its container. Stronger concentrations of peroxide are used by professional bone cleaners; however, these concentrations are quite dangerous and require special handling, storage, and personal protective equipment (PPE). Nitrile or latex gloves and PPE should be worn when handling any peroxide as it can cause chemical burns. All peroxide can cause chemical burns to the skin and react violently with metals. Every Vulture should take the time to educate themselves with the proper PPE and handling techniques of hydrogen peroxide and any other chemical they plan to work with.

Other chemicals are often recommended for whitening bone. A beauty care product called developer is a crème like solution of peroxide, alcohol, acid, and other ingredients used to help protect human hair during the whitening process. It is not recommended to use this type of product to whiten bones, as it does not reach all the tiny recesses of a skull and is not really suited to the application. Chlorine bleach should never be used on bone. Chlorine bleach will cause flaking, yellowing, and ultimately the complete destruction of your specimen.

Transfer your specimen, including all teeth and small bone pieces, into a clean glass or plastic container and cover in peroxide. NEVER use a metal container, and never seal a container with peroxide. It is okay to cover the container but do not seal it tightly. Once your specimen is completely covered in peroxide you will notice bubbles forming, this is ideal! Peroxide releases oxygen gas ($O2$) when it breaks down during the whitening process; this gas should be allowed to escape freely.

Remember that black color discussed earlier? If you have this develop splash a little peroxide on it and watch it disappear like magic! If your specimen begins to float use a couple layers of paper towels to keep it completely covered. The paper towels act as a wick and draw the peroxide to the exposed portions of the skull. This technique can be extremely useful when cleaning skulls with antlers. You can use paper towels or even white fabric rags to wick the peroxide onto the pedicle and help prevent color loss on the antlers during whitening. Be sure to use a good strong paper towel that doesn't break down quickly.

The whitening process can take from a few hours to several days. Room temperature is best for whitening, 70 to 90 degrees Fahrenheit. Keeping peroxide warm is a challenge. We recommend using a sunny window or a heated mat for safety; rather than a submersible heater like an aquarium heater. Waterproof heated mats designed for plants or animals have an imbedded heating element that is controlled by a thermostat. Just set the containers on it and cover to keep the heat in. An old cooler works well for covering small containers. Small specimens can also be whitened in a clear container in a warm, sunny location. Remember to keep pets and animals away from all peroxide containers.

Once the specimen is to the level of white you desire, drain off the peroxide into a clean container, again being careful not to lose any teeth or small parts as the peroxide can make them float! Peroxide can be reused until it goes "flat" and no longer foams when you put bones in it. Store peroxide per the recommendation on the bottle. After the specimen is fully drained it should be rinsed in warm water, allowing it to soak for five to thirty minutes to ensure peroxide is rinsed from all cervices.

Specimens should then be transferred to a tray or shallow tub to dry, out of direct sun, away from heat sources and away from fans or moving air. They should be dried slowly, to help prevent warping and cracking. Rotate them several times over the course of the first few days to keep them drying evenly; if they dry too slowly specimens are at risk of mold or mildew. If a specimen begins to show mold or mildew, go back to the maceration process and repeat all the steps. During the drying process you may notice the teeth beginning to crack. This is natural, unavoidable and unpreventable. Slow drying and keeping skulls at a consistent humidity will reduce cracking. Canines and teeth with two roots are most prone to cracks. If a tooth cracks, don't worry, I will discuss repairing them later in the process.

Once a skull is dry, it should be allowed to sit at room temperature. This will show if any grease or smell is still present. If your skull begins to bleed grease, just start the degreasing process over. Grease shows up as spots on the specimen that may look wet or oily. Over time these spots turn to yellow or orange, in extreme cases they may even mold!

Maceration is the easiest and least expensive option available to beginner Vultures. More experienced Vultures, schools, and private collectors also use dermestid beetles, wild insects, and oxidation for flesh removal. Vultures in rural locations may also employ nature cleaning or burying for flesh removal.

Dermestid beetles are an efficient method of flesh removal from large and small specimens. Their consumption of soft tissues can be carefully managed to allow specific ligaments, tendons, and cartilaginous tissues to remain. This is often done on very small specimens that will be articulated. Referred to as a ligament articulation, this technique allows the tiny bones of the feet, ribs, and tail to remain held together in a natural position. These ligaments are rehydrated and then the specimen is carefully pinned while it dries. Few pins and little glue are used in ligament articulations. Beetles are also effective for fetal or very young specimens but must be carefully managed, so they do not damage or burrow into the extremely soft, underdeveloped bone.

Beetles are not for the beginner. Dermestids require very specific conditions and habitat management. They require temperature and humidity-controlled enclosures. Dermestids are prone to mite and maggot infestations that can devastate a colony. Colonies take months to grow and starter colonies available online may take a week or more to clean even a small skull. Healthy colonies used by commercial bone cleaners can have hundreds of thousands, or even millions of beetles. Dermestids must be fed, even if no specimens are to be cleaned. They are perfectly suited to the consumption of dry natural tissues; however, this means they can be exceptionally destructive in a home or business setting. These carrion beetles will not only eat the dried tissue of your specimens, but if allowed to escape they will also damage taxidermy, hides, furs, horns, claws, insects and mummified specimens found in your collections. Left to their natural habits they will also begin to burrow into the sheetrock of your home. They are extremely hard to eradicate from any dwelling. It is best to keep them in their own well managed building away from your home, business or collections.

Beginner Vultures may wish to experiment with wild insects for flesh removal. These include carrion type beetles and the larva of several species of flies, commonly called maggots. Maggots are by far the easiest insect to obtain for the purpose of soft tissue removal. A container with a small hole containing a hydrated, fresh specimen left outside will quickly attract local flies who will lay eggs on the specimen. These eggs hatch quickly and the larvae immediately begin digesting the tissues. Maggots are most effective in warm temperatures (>70 degrees Fahrenheit); they quickly remove soft tissues but are less likely to consume tougher connective tissues like tendons and cartilage. Care should be taken with maggots to keep them from pupating in the nasal cavity and tiny cervices of a specimen as the pupa cases are nearly impossible to remove. If it appears the maggots are reaching the pupa stage the specimen should be flooded with plain water to encourage the maggots to leave the skull. It can take just days in warm temperatures for fly eggs to hatch, complete the larval stage and pupa. Cooler temperatures will slow the process to several weeks. They do not need to go to waste though; maggots make a good protein source for backyard poultry or as fishing bait. Care should be taken to destroy all maggots before pupation to avoid an explosion in your local fly population.

An alternative method for small, but durable specimens is oxidation. It uses the natural properties of hydrogen peroxide to soften and facilitate the removal of small amounts of soft tissues. This can create a specimen suitable for a ligament articulation, or a finished skull. Specimens to be cleaned by oxidation must have the brain and as much soft tissue as possible removed before the process begins and must be kept fresh as a spoiled specimen will not hold together during the process. Keep a specimen frozen until you are ready process.

Another technique employed by some bone cleaners is burial. The success of this method varies greatly with the local soil and climate. Very dry or very porous soils that drain too well will mummify a specimen before the soft tissue is removed. Damp acidic soils will preserve the soft tissues, much like the bog mummies found in Europe or begin to saponify a specimen, turning its fats to adipocere. The soil in some locations will actually remove the flesh and then begin to dissolve the bone itself. Without a degree in soil science and lots of testing, it is just a guess and check to see what type of soil you have. Start small and see how it goes. One commonly accepted method for disposing of dead livestock is to compost the remains in a manure pile. This is extremely effective but still results in bone damage. Soils also tend to discolor bone and it is unlikely than any buried specimen will whiten to the level of a macerated or beetled one.

Several tissue removal methods discussed in online forums and informational sites should be avoided. These include but are not limited to boiling, simmering, acid baths, pressure washers and pressure cookers. All of these methods present a very high risk of damage to the specimen and should be avoided at all costs.

Regardless of the method of soft tissue removal used on a specimen, the process of degreasing and whitening remains the same.

Reassembling Skulls

Once you are certain your specimen is finished to your preference, you can reassemble it. In most species the lower jaw will separate during the cleaning process, most teeth will fall out and the upper portion of the skull will remain intact. Some species, like deer and sheep, will separate at the nasals and premaxilla bones. The premaxilla bones should be glued on first, followed by the nasal bones. Very young specimens will have unfused sutures, growth lines in the bone; these sutures will separate during most cleaning processes, leaving your skull in many pieces. Recently coined "skuzzles," these skull puzzles can be challenging to assemble. Extremely young specimens may be in as many as 40 pieces. Care must be taken to avoid losing any tiny parts during cleaning and reassembly. Oddly enough, opossums tend to fall apart at any age.

The next page has a photo showing the various parts of the "skuzzle". The first step is to find the two pieces of the skull that look like birds in flight. One has taller wings, the other is flatter and less bird like. You want the taller one. This piece, and the two largest pieces of the top of the skull are your foundation for building your skull. Glue these three pieces together first. Bone should be slightly damp when assembly begins to avoid warping or damage to the delicate bone. Then assemble the bones of the face (either side) and the bones of the ear, separate from your center piece. Allow these to dry. Then assemble the bones of the back of the skull, with the ear pieces. Rubber bands will help you hold this together while the glue dries. For very delicate things you may have to hold the bones by hand while they dry. After assembling the back of the skull, you will assemble the two front halves of the skull while balancing the largest of the delicate sinus bones inside. This piece has a long U-shaped ridge with two pyramidal honeycomb sections. Once that glue has dried you will glue the smaller pair of the sinus bones into the nasal cavity. The pointy end goes towards the brain and the side with the flat edge goes toward the outside of the skull. Then glue on the flat upper nasal bones. All assembly should be done with white Elmer's Glue. White glue, patience and rubber bands are extremely helpful when reassembling juvenile skulls.

Extremely delicate bones may be strengthened by using an acrylic resin called Paraloid. This soaks into the surface of the bone and bonds with the cell structure. More about this technique can be found online.

When preparing to assemble the skull of a mature specimen, you will need white Elmer's type school glue, a few rubber bands, and a pair of tweezers may be helpful. It is best to first lay out your pieces on a clean work space. Carefully test fit each tooth until you find the socket it belongs in. If any of your teeth have split, you will need to match up the pieces. Broken teeth should be glued back together when they are in the socket. Otherwise they can rebreak during assembly or change shape and no longer fit in their socket.

A few hints for replacing the teeth of most North American Furbearers:

- Top canines are straighter; bottom canines are more curved. The groove goes to the inside of the mouth.
- Upper and lower incisors are single rooted. Upper incisors have shorter roots than lower incisors.
- When assembling incisors, the short side of the crown goes toward the canine tooth.

Tips for most Herbivores:

- Incisors have a flat side and a pointed side. Flat side goes towards the center of the jaw.
- Largest pair of teeth go in the center of the jaw.
- Molars in herbivores typically do not come out and if they do they are very easy to match to their correct location.

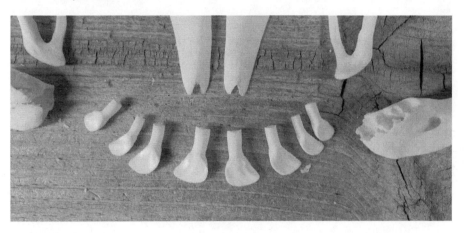

It can be helpful to match up pairs of teeth, especially for premolars and molars. Make sure all your teeth fit into the skull before gluing any in. After you are sure of position, carefully remove the loose teeth and arrange them in their corresponding location around the specimen. Carefully put a small amount of glue in the socket, be careful not to get any on the visible portions of the skull, then reinsert the teeth. Wipe off any excess glue. White glue is great, it dries nearly clear and will not yellow over time. If you ever decide to degrease or antique/paint a skull, white glue will dissolve in water. Super glue should be avoided, as many brands will yellow over time. Also, super glue cannot be removed if you need to degrease or decide to paint or antique a specimen.

If a specimen is very tiny, a hypodermic needle/syringe can be very useful for applying glue. It is a good idea to carefully cut the tip of the needle off for safety and control. An 18 gauge or larger needle works best. After all of your teeth are in and your glue is dried, apply glue to the two halves of the lower jaw, use the rubber band to keep the halves together and articulate the jaw to the skull. If you allow the jaw to dry not articulated to the skull you run the risk of it not fitting. Check that your jaw halves are level and allow to dry. If you find your jaw wants to move side to side, you can use a toothpick to keep the halves at the right distance. Once the glue has fully dried you can remove the rubber band and the specimen is ready to display!

Horn sheaths sometimes shrink during drying and if this occurs, one of two methods can be employed to properly reassemble the skull. If you do not need to remove the horns once a skull is finished you can soak the horn for 30-90 minutes in HOT tap water to soften it and then reassemble. As the horn dries it will shrink on to the core and become permanently fixed. If you need to be able to remove the horns from a finished skull (for shipping or educational display) a Dremel or sander can be used to reduce the diameter of the bony core. Remove material slowly and test fit often. Typically, with this method you will need to drill the back side of the horn and insert a small pin or screw to secure the horn onto the skull. One notable North American species grows a horn that cannot be moved. Musk ox, *Ovibos moschatus*, grows a horn with a thick base that curls down towards the lower jaw and then tips back up. The shape of these horns makes them non-removable and difficult to clean. When preparing a skull like this it is best to keep the entire skull and horns submerged and complete the process as quickly as possible to avoid any damage to the skull.

Mature antlers require little attention before cleaning a skull. However, after processing, you may wish to enhance the color of the antler or repair any fading caused by the cleaning process. Several products are commercially available for recoloring antlers. Potassium permanganate, wood stains, acrylic paints and natural dyes, can all be used to restore a natural color to antlers. Apply any color in light layers and build up the depth of the color and texture. Applying one or two heavy coats will result in an unnatural look. Fine sandpaper, steel wool and even branches can all be used to rub on an antler and bring out a natural appearance.

Long-Term Care

Properly cleaned skeletal specimens will offer beauty, education, and reference for many lifetimes. They will, however, become dusty over time. It is best to clean bone with compressed air as it will help get any settled dust out of the porous bone surface. In extreme cases; it may be desirable to re-whiten a specimen. This should be avoided with very old or delicate specimens as they can be damaged. As bone ages the bond between the calcium phosphate and collagen of bone will weaken. Adding water will essentially unravel your specimen causing extreme brittleness or complete disintegration. Without water, this cannot happen. Do not attempt to re-whiten any specimen if you do not know its full history. Specimens should be stored out of direct sunlight and away from any heat source as these can cause significant damage in the long term. If they need to be packed away long-term, skeletal specimens should be wrapped in paper and stored in a breathable container away from high humidity, temperature swings and pets. Horns and hooves should be stored in a separate, sealed container with mothballs or naphthalene to combat insect damage.

In ancient times, it is believed bones for cultural use would either be nature cleaned by propping in a tree or place away from scavengers, or by cooking and removing the tissues. Beginning in the very early 18th Century, arsenic was used by museums to preserve skulls and skeletons and protect their collections from insect damage. Fortunately, this practice fell out of favor in the early 20th century. However, it is still possible to encounter arsenic preserved specimens at flea markets and second-hand shops. Any vintage specimen should be treated with care when cleaning, dusting, or handling.

Vulture Culture can be rewarding for young and old alike, but please remember to wear the appropriate PPE, be aware of neighbors and familiarize yourself with the MSDS of any chemical you choose to use. Stay safe, wash your hands… and yes, Febreze will take that smell out of your shoes!

Skeleton Articulation[19]
By Amy Wilkinson

DISCLAIMER: I am in the UK where the Migratory Bird Treaty Act does not apply. Some of the species I mention here are not legal to possess in the United States; good alternatives include species which were introduced to the US like the European starling and house sparrow, or domestic birds like chickens and pigeons. Please make sure to abide by your local laws and legislation!

My first articulation was a red legged partridge – the first of its species I found. Sadly it had been laying dead by the road for several days. Its lovely plumage was caked with dirt, and its feathers fell out at the slightest touch – too far gone for taxidermy! But I still wanted to honour it in some fashion, so I decided I would articulate its bones standing as it would have done in life. I found it fascinating to see what was inside of this animal when it was alive, how each bone fit just so with its neighbour. It was the perfect project to complete with an animal that was too rotten for taxidermy or any other method of preservation.

Articulation is the process of putting together the bones of an animal in a pose which it would have held in life. Skeletal articulations were used primarily as teaching aids, with animals like rabbits, pigeons and chickens in a sitting or laying down pose kept under glass covers in classrooms. They are popular now as art pieces, with animals in more elaborate poses. With a Dremel or similar tool and some imagination, there isn't much limit to what you can do with a skeleton!

---

[19] Please note that this is not meant to be more than an introduction to skeleton articulation; it's a very complicated process that can vary from species to species. Consider this a good set of tips for articulations in general rather than a detailed step-by-step; there are book references the essayist gives later that are excellent how-tos on specific species.

Finding a Skeleton

The first thing is to get yourself a skeleton to articulate! You can buy an already cleaned one, or clean your own. If you want to buy a skeleton, online is generally going to be your best source. General sales sites like Etsy and eBay will likely have some. If you are part of vulture culture sales groups elsewhere and comfortable buying privately using Paypal you will likely come across species that Etsy and eBay do not allow (such as domestic dogs and cats). If you want to clean your own skeleton, refer to the bone cleaning section of this book!

There are some animals which are easier than others for a first articulation. I class things as "easier" in that their bones are of a manageable size to work with – generally large enough to glue and wire, but not so large that the bones are heavy enough to require mostly wire. So it's not terrible if you only have a "difficult" species to work with – it just means a different approach when putting the skeleton together.

Easy species to start with are small/medium sized birds, mammals and reptiles Rooks and crows tend to be easy to obtain and aren't difficult to start with. Pigeons, blackbirds, and other similar sized birds are good. Chickens and pheasants may be larger but are also easy to obtain. Smaller birds such as tits, robins and the like are trickier due to their smaller and more delicate bones, though they are easier to articulate with only glue.

With mammals, species like squirrels and rabbits are good in terms of size. Foxes and raccoons are relatively easy to obtain though slightly larger to pose and fix in place. Though not as easy to ethically source and a bit controversial by some people's standards, cats are a good size for a skeletal articulation. Doing a partial articulation of skeletal sections from larger animals, like pig, deer and cow limbs, makes for good practice as well.

Preparation

As mentioned earlier, you may be starting with a carcass that needs to be processed.

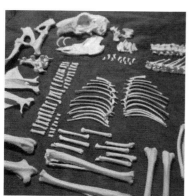

When cleaning a skeleton that you intend to later articulate, the main thing is to keep all of the animal's bones. This is difficult with small toe bones and teeth as they can become easily parted from the rest of the skeleton. If you have a whole animal which you know you wish to articulate, there are some things you can do to make the articulation easier for yourself.

Maceration involves letting the flesh of the animal rot in a (usually sealed) container of water. It's a contained method of cleaning bones, and by using a sieve when changing the water you can ensure that you do not lose any small bones. Please see the tutorial on bone cleaning in this book for further directions and tips.

It is easier to separate the limb bones into different containers before cleaning the skeleton. Label the containers, and when it comes to laying out the skeleton you will know for sure which pile of toe bones are from the front right leg, and which are from the back left leg!

Once you have your skeleton (whether bought or self-cleaned) it's a good idea to lay it out on a flat surface. This gives you a chance to become acquainted with the skeleton - and if it's missing any bones! It also gives you time to look at the slight differences between each bone and where they all go. It's easier to figure this out on a flat surface than midway through an articulation. If you don't have space, or don't want to lay out the whole skeleton beforehand, it is still very useful to separate bones into groups or pairs, so that they're easier to find whilst fixing things together, for example keeping the ribs in their respective pairs instead of one big pile. Predetermining foot bones will save a lot of time and stress later on – the bones are very similar with only minimal differences dictating where each bone sits. Here are a few things to keep in mind:

- It's an interesting part of nature that most animals are bilaterally symmetrical – if you cut them in half from nose to tail both halves (should) be the same. Use this fact to your advantage! The curve of toe bones from the left side of the body will curve in the opposite direction to the toe bones from the right side of the body. If in doubt, pair bones with have the same shape and curve. At worst you will simply have the pairs inverted (left on the right, and the right on the left).

- Pairs continued: as all bones that are paired are paired laterally, it also means if you have an odd number of bones, you are likely missing one.

- Baby bones: in the case of juvenile animals which had not finished growing before they died, they will have more individual pieces of bone which make up their skeleton. This is due to a large number of bones actually starting as several pieces of bones attached together with cartilage, and as the animal ages, the cartilage turns to bone, fusing the bones into one solid bone. Some animals take far longer for their growth plates to fuse until they are no longer visible. Some never completely disappear, such as the sutures on a skull, or the growth plates on this deer leg bone.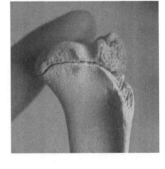

- Broken bones: in the case of animals which are found by the roadside, it is not uncommon for them to have one or more broken bones. Look for the sharp edge, and then find a similar bone with a corresponding sharp edge, though sometimes bones are crushed rather than broken cleanly. In the case of a clean break, I just glue the parts back together. This is often true of the skulls of roadkill animals, which can be a jigsaw puzzle in and of itself. With bones that are more severely broken, you can use the bone as it is, with a wire linking it in place. Alternatively, you can just remove the broken bone(s) and replace them.

- Missing bones: there are several options if you find you are missing a bone. If it's a small bone at the end of a limb, then you can simply articulate without it. Either ignore the fact it's missing and articulate its neighbours together, or leave a gap to recognise that one bone is missing. If it's a major bone like a humerus or fibia, or you don't want to articulate with bones missing, you can source the bone that's missing from another skeleton. This works best with an individual of the same species (and ideally the same age) as it will fit best. You could also make a replacement bone from clay, epoxy putty or wood.

Strike a Pose

Planning out the skeleton's pose and looking at references is best done before fixing bones together as it's difficult to change something when half the bones are already glued and wired together. The main problem is making sure you can attach the bones together securely and finding references for how the bones attach.

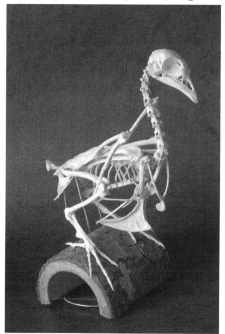

The easiest pose is generally with the animal standing in a relaxed pose – this spreads the weight across all limbs. More dynamic poses involve a bit more planning in how the weight of the skeleton will be spread – whether through the other limbs which are in contact with the ground, or through a main wire which supports the whole animal. You're also not constrained by having the animal stand; you can mount it in a leaping pose hanging from ceiling for example, or lounging over the edge of a shelf. The pose is only limited by the tools you have, your imagination, and putting the skeleton together!

There are more references for static poses than dynamic, and any dynamic pose will need stronger attachments for the bones to keep everything in place since weight is not as evenly distributed. Tipping points are also important for any skeletons which are articulated without a base.

A tipping point is when the weight is so far out or over from the contact point of the skeleton against the ground, that the whole articulation falls over. Such as with a game of Jenga when too many blocks are missing from one side of the tower, the combined weight will pull the whole assembly over.

A base gives the whole skeleton a weight which prevents it from tipping over, as well as being easier to move and protecting the whole articulation. Dynamic poses where the skeleton has only one limb in contact with the ground require a large base to offset the rest of the skeleton's weight, or carefully concealed weights in a smaller base. For good references, check out the taxidermy mounts of big cats chasing prey, and pheasant fighting; you can look at photographs of live animals as well. There will be more variety in poses, but it will require some more thought into how you can replicate the pose.

With poses in which the articulation will hang free, such as from the ceiling, you will have to think about the attachment point(s) and where the rest of the weight of the skeleton will sit. If you have only one attachment point, unless it is perfectly balanced your skeleton will end up hanging with its head or tail pointing up. Ideally you need two attachment points, one near the front and one near the back of the animal (this prevents tipping front and backwards). You also should be mindful of the articulations pose and balance to stop the skeleton from rolling from side to side.

The weight of the skeleton (the bones themselves and the materials that you use to articulate it) is best kept low, and in the middle of the skeleton. As previously mentioned, the easiest pose to balance (with regards to the weight of the articulation) is a four limbed animal standing with its weight equally split between each leg.

Something to keep in mind when planning your pose is if you will be using wire, where will it go and how visible will it be? Dynamic poses will require a continuous thick gauge wire anchored firmly in an overly large base (to counter the weight of the skeleton and prevent it tipping over). In taxidermy this wire would be hidden within the forms and stuffing used, but you have less material to hide it in with skeletons! Some options include drilling the bones so you can thread the wire through the middle, or discreetly hiding the wire along the underside of a bone. Drilling the bones to thread the wire inside will produce the most seamless results, but it is the most difficult and time consuming to do. Depending on the size and shape of the bone, it can be difficult to drill. When first starting out, hiding the wire alongside the other bones is the easiest method.

Another point you can consider whilst planning your pose is where you will start your articulation. This ties in because if you plan a very complicated pose with many hidden parts, you may find if you have gotten carried away articulating the feet first, that the angles at which you've glued the feet bones have caused a major change in the angle the rest of the skeleton stands at. The spine is nearly always the easiest place to start with as it is the meeting point for all other parts of the body. But again, this is not true for all articulations!

References

This is a separate section because of its importance. You can never have too many references! As mentioned above, you can look at live animals and taxidermy mounts for an idea of how to pose your skeleton, but the main use of references when articulating is identifying which bone is what, and where it goes. In theory, you can use any photo of an already articulated skeleton. But this has its pitfalls – the person who articulated the skeleton may not have used references themselves, and thus bones may be in the wrong place. The best reference material would be MRI scans or x-rays of living animals, where all the bones are still held in their correct places by flesh and tendons. Sadly these are often difficult to find online! Even if you cannot find your EXACT animal, try using a reference photo of a similar one, e.g. using a cat skeleton to reference for a civet articulation, or a chicken skeleton for a duck.

Some museums have provided online resources of their skeletons, including this amazingly useful 3D model of a dog skeleton: http://www.real3danatomy.com/bones/dog-skeleton-3d.html

If you prefer comprehensive paper references, here are some suggestions for good books:

- *The Bone Building Books* by Lee Post: all brilliant books on how to articulate difference species.
- *Mammals of Britain* by M. J. Lawrence and R. W. Brown: an old book, but comprehensive on UK native species. Includes illustrations of different individual bones compared to different species.
- *Anatomy Drawing School* by Geza Feher and Andras Szunyoghy: I have the 2006 edition, which has some great references of horse, dog, cat, sheep, cow and a couple of other animal species. It shows illustrations of both complete skeletons and close ups of individual bones and joints.
- *Introduction to Veterinary Anatomy and Physiology Textbook* by Victoria Aspinall: though more biology than skeleton specific it does have some nice references and informative tidbits
- *The Unfeathered Bird* by Katrina van Grouw: a book of illustrations of just bird skeletons

Tools

There are numerous different ways of articulating skeletons. Here I will go over the pros and cons of each, and recommend what they're best for. What I find works for myself may not work for you, and you may find similar or alternate tools which work better for you.

- Glue: depending on the glue, it can be quick or slow setting. Glue is a good method of attaching small lightweight bones to each other. It is only as secure as the strength of the glue, and the number of contact points between the two bones you are trying to stick together. I tend to use glue on the end of limb bones which do not bear any weight, and my go-to-glue is standard super glue. Other glues I have seen recommended are gel super glue (which tends to not run and make as much of a mess as standard super glue), PVA/white glue (which isn't as strong but will still suffice with small bones) and Gorilla Glue. All should be easily available from any decent store.

- Wire: best for weight bearing, ie holding the limbs in position and attaching the skeleton to a base. There are different gauges of wire, with thicker wiring able to bear greater weight. Around 3mm is good for small-medium sized articulations, and if you can't find it in a local crafting or DIY store it can be purchased online.
- Movable joints: articulations have been done using springs, wire and metal hinges to create a skeleton that is one complete animal, but with posable joints. This is not commonly done, as it would be very time consuming and difficult to do. I have considered that magnets could also be used as an attachment for bones, with the joints drilled and magnets glued inside for a hidden joint. I include this as a note that you don't have to use what is recommended as a project can be as simple or as complicated as you want to make it.
- Base: this can be made out of any material, but it's commonly made from wood. eBay is often a good place to shop for a base for an articulation as are craft shops, but you can make your own if you want to.
- Zipper storage bags: invaluable for keeping different bones separate. Especially useful if you aren't able to articulate your skeleton all in one go.
- Tweezers: for handling and placing small bones
- Magnifying glass/lens: for better visibility when handling and placing small bones
- Pins and polystyrene: invaluable for keeping bones in place whilst they dry. You can also use it as a temporary stand for the full articulation if you've decided against using a base.
- A small drill: either a hand drill or small motorised drill like a Dremel. A necessary piece of kit if you plan on drilling any of the bones you will be articulating! Don't forget eye protection and a suitable dust mask when drilling bones – the resulting dust is not kind to human lungs.

Other things you will also find useful: A strong, bright light source and a nice flat surface on which to work. Natural light will show any discolouration better than most artificial lights, but any strong lamp will work. I like to have my references on hand, and tend to keep my laptop nearby in case I want to look up additional references in the midst of articulating. It's also a good idea to keep some acetone (nail polish remover general works as a lower concentration of acetone) nearby when working with superglue, in case you stick yourself to the bones! In that same vein, rag cloths and/or kitchen rolls are handy if you spill anything, or quickly need to remove excess glue. To remove PVA/white glue, you simply need to soak it in warm water and it will come apart. Glue gun glue can be picked off once it has cooled and hardened.

Articulation

Once you have all your tools, your skeleton and your references, you can start building your articulated skeleton. Remember this fact during your articulation: with any healthy bone the individual bones tend to fit together when in the right place.

Where to start? First is fixing any broken bones so that you have complete bones to articulate. As mentioned earlier, it is usually easiest to start with the spine as it carries the main weight of the body and conveys a lot of movement and emotion. Then look at the base / anchor point / attachment. In the case of poses which sit/stand, limb bones next. But if you have your pose planned out exactly, you can always start with more intricate parts of the articulation like paws.

The back bone of the articulation: Remember how (healthy) bones fit together? This is true of vertebra though the differences between each of the bones are slight which makes initial placement difficult. But if you have the bones already grouped into similar shapes, and a reference photo or two, you can piece together which bone goes exactly where. With vertebra most land animals have very similar bones: 7 neck vertebrae, 12 thoracic vertebrae, 5 lumbar vertebrae and then the tail. The neck vertebrae tend to be compressed to take the weight of the skull. The thoracic vertebrae are split into two groups - the chest vertebra which act as the anchor points for the ribs and the muscles which attach the front limbs, and the lower back vertebrate which support the lower abdomen muscles and organs, and attach to the pelvis.

I tend to thread my vertebrae onto a larger, strong gauge of wire using the naturally occurring hole where the spinal nerves lay in life. Adhesive between each vertebrae will stop them wandering, but the best method of attachment is drilling through the core of each vertebrae and threading the wire through the drill holes. Tails are extensions of the spine and much the same to assemble. They and the ribs are often awkward to assemble in small/medium species because they may not be large enough to be drilled and have wire threaded through, so adhesive is the primary fixative instead.

If you are patient and steady handed you could partial drill both the ribs and vertebrae, and use small pieces of wire AND adhesive to fix the bones in place. This is the most secure method of attachment, but also the most time consuming. If you have clamps and a stand for your Dremel it would be quicker than attempting it freehanded. The downside to the most secure method of attachment (wire and glue) is that you are making permanent changes to the bones, and it is difficult to correct things if they go awry. This means you have to keep in mind the overall pose whilst drilling and affixing each bone, otherwise the articulation may look off, or worse, unbalanced.

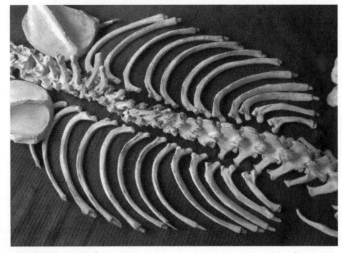

In real life the curves of an animal's body are naturally smooth, with few right angles or abrupt corners. A spine usually curves from the front to the rear of the animal, and the ribs then curve out from the spine, whilst also tapering.

From the spine then working down to the tail, I ignore the skull till the very end, as it's generally a heavy and cumbersome attachment, so best fixed on once the rest of the skeleton has been assembled. But do make sure to leave yourself enough wire protruding from the skull end of the spine that you can use it as an anchor point for the skull! In larger species the skull has a considerable weight all of its own, and this will require a firm attachment to the rest of the articulation to prevent it falling off at a later date. Small species are easier to attach the skull – a small amount of glue and wire usually suffices. If you want to orchestrate it, it is possible to attach the skull so it may be removed at a later date if you choose. Some people like to attach the jawbones so that they are hinged to the skull as in real life – again, you can make it as complicated as you choose to!

The pelvis is an interesting bone as it attaches to several different bones, including those in the legs and tail as well as the spine. In juvenile animals it's made up of four (sometimes more) pieces of bone, but in most adult animals the pelvis will have fused into one bone. Bird pelvises are very strange looking, and often mistake for skulls. Mammalian pelvises are all very similar with a distinctive shape. They can be difficult to orient, and at first, you'll want to spend time looking at your references and making sure you have the pelvis in the right way before attaching it.

As the spine and ribs form the core of most articulations, it is often worth taking time to look at the overall piece once you have finished attaching the spine, pelvis and ribs together. If bones are off kilter, or the pose is unbalanced, it will be easiest to rectify before you begin attaching anything else.

The legs of the matter: In animals which walk on four legs, the front limbs are awkward because it's not strictly attached to the spine in most species. Rather it "floats", tethered by muscle in life. This means there is no handy attachment for the front leg, other than what you create. The method of attachment for the front limb is then dictated by the base/attachment for the whole skeleton, and if the front limb is actually weight bearing.

This is where the pose planning comes into play. With the ribs already fixed, you will have to be careful applying pressure to attach any other bones. But the ribs lay under the limbs, so they need to be attached before the limbs. The easiest pose is often one which anchors the front legs by another means than to the spine (but that doesn't mean it's impossible to articulate in a different pose).

So, in the case of animals with significant front limbs (like foxes, cats etc) I attach the shoulder blade to the spine and ribcage first. I then build the rest of the front limb. This gives you a point at with you can attach the rest of the limb, without wrestling with the whole articulation. With smaller species, (amphibians, reptiles, birds) that are posed in place, you can build the limb in place as you'd likely be using glue.

Building the limbs is one of the easier parts of an articulation, with difficulty varying between species. Limbs have four parts to them - the attachment/anchor point to the spine, the top of the limb, the bottom of the

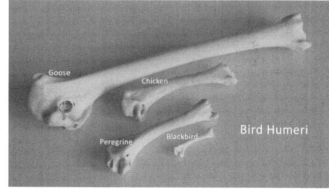

limb, and the foot of the limb. All bar the foot bones are large and easy to identify. Even between species the bones are identifiable (see the picture comparing the same bone in different species of birds). As the differences between the limb bones are quite distinct I won't go into them in much detail here. The worst part of any animal to articulate is the feet. All animals have a number of small bones in the feet, which in life slot neatly together to provide a point through which the animal's weight is transferred to the ground. After death and cleaning, these bones are difficult to sort between due to their number, (generally) small size and slight differences. In a healthy fox you'd have four paws made of at least 28 bones in each paw, which makes a total of 112 bones! However, some species are easier than others. Mammals with five toes are the worst, like, absolutely terrible, but those with less than five toes (such as pigs, cows, goats and horses) are easier as they have less bones in their feet (see the pictures of the dog and ungulate feet on the next page). Birds are easier again as they have very simple attachments from toes to the base of the legs. But for the majority of species you'll be sorting through the small bones of the wrist and ankle. In life these form a flexible attachment from the base of the leg to the toes.

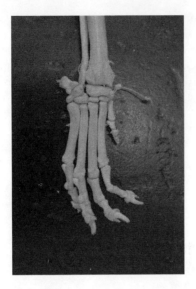 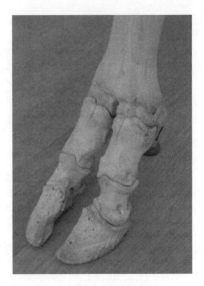

Once the feet are assembled it becomes a matter of attaching the skull (unless you did it earlier when assembling the spine). If you have your articulation free standing, you are done! Congratulations on your first articulation, and good luck with your future projects!

Mouse Taxidermy
By Divya Anantharaman

Getting started with taxidermy can be really intimidating. Though mice are small and fairly delicate, working on a mouse for your first project can be a great starting place because you'll be introduced to all the steps of the taxidermy process in a small size. A small specimen also requires less working space and storage space, and no matter where you live, most people are familiar with mice, making it easier to find good reference images for making your mount lifelike! Mice are also quite easy to source, whether you use frozen feeder mice sold as food for carnivorous pets (available at specialty pet shops and through online suppliers), naturally deceased or roadkill finds, or deceased domestic specimens (obtained by building a sensitive relationship with pet breeders or nature centers.) When collecting roadkill, you should be sure to research your local laws or check with government wildlife officials--in some places, even common species are protected!
Before you start,

Thaw your mouse in denatured alcohol (or 90% isopropyl alcohol if you can't access denatured) to help toughen the tissue and prevent slippage (when the top layer of the skin and hair slips off due to bacterial damage, warmth, or other factors.) This will not reverse any damage already done, but it does help prevent new damage! Thawing in water can harbor bacteria and cause slippage, and water will not have the same tightening effect on the skin that will make it easier.

Supplies

For your first taxidermy project, you may find that you can use supplies you may already have in your workshop, art studio, or craft corner. This tutorial tries not to use too many speciality tools, and most can be bought at art supply stores, hardware stores, or at major online retailers. If you are just getting started, use good tools (a sharp knife is a safe knife!), but there is no need to purchase anything too expensive! Aside from your mouse, you will need:

- Reference photos - you can find photos of live animals on the internet doing nearly anything! Having a file for any animal you are working on is great, and printouts of the photos that apply to your current project will be very helpful in reminding you of all the details. Anthropomorphic poses may require a little more creativity and exaggerating on what a mouse can do, but natural reference is still helpful in sculpting your form and building an understanding of what the specimen's body can do)
- Nitrile gloves to wear while skinning and fleshing
- Work surface - can be a dissection tray, non-porous countertop, or sheet of plastic
- Cleaning supplies for your work area and tools (paper towels, sponges, disinfectant)
- Dawn or other neutral degreasing dish soap
- Cup or jar to thaw mouse and soak skin
- Denatured alcohol or 91% Isopropyl alcohol
- Borax or taxidermy grade dry preserve powder, or small mammal tanning kit (we used borax here)
- Scalpel with #11 blade or #22 blade, or a hobby knife (i.e. and Exacto or Excel knife) for skinning
- Curved cuticle scissors
- Curved dull blade or small butter knife for fleshing
- Soft wire brush for fleshing
- Fleshing beam (any surface to stretch the skin over so you can scrape the skin. We used a wooden egg from a craft store and a fleshing horn. You can use a piece of wood, the end of a baseball bat, scrap piece of PVC pipe, or anything you have laying around!)
- Carving foam to make body form (available at sculpture supply shops, or through taxidermy suppliers. We used a foam from a taxidermy supplier here, but you can also use the rigid blue foam from Dow or balsa foam. It is not advisable to use floral foam or white styrofoam because neither are strong enough.)
- 18 g stainless steel/non rusting wire
- Pipe cleaner
- Scissors

- Air dry clay (Critter Clay is great for low shrinkage!)
- Apoxie Sculpt or Magic Sculpt (two-part "epoxy clays" that have minimal shrinkage and dry fast)
- Carving tools (knife, rasp, sandpaper)
- Modeling tools/dental tools
- Eyes (we used black glass head round pins, you can also use beads or specialty eyes)
- Thin, sharp sewing needle (a dressmakers needle is used here)
- Synthetic thread (Poly/Nylon blend, or lightweight fire line)
- Super glue
- Latex caulk
- Disposable syringes with and without needles
- Paintbrushes
- Paint and/or powdered pigments for adding color to faded parts after drying (powdered makeup can work here too!)
- Foam or wood mounting base
- Masking tape
- Dressmakers pins
- Insect pins
- Blowdryer
- Soft toothbrush for grooming fur
- Any props or accessories you would like to use!

Skinning

Our goal is to remove the skin from the carcass and keep both intact. If either are damaged, don't get discouraged! You can always make repairs. The method used here is called dorsal skinning--all that means is the incision is made down the back. The fur and skin are thickest back here, making everything easier to handle.

1. Use the dull side of the scalpel to part the fur from where the head and neck meet to about one centimeter from the tail. Hold the scalpel like a pencil and use the tip of the blade to trace this line, cutting just through the skin. An alternative method of making the incision is by holding the scalpel blade facing away from you, and sliding it up from the underside of the skin. Either method will prevent cutting through too much fur.

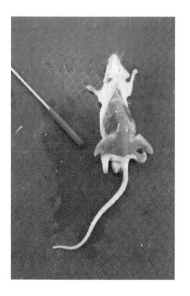

2. Start gently peeling the skin away from the body. The motion is similar to how you'd peel an orange. Slide and wiggle your finger between the carcass and the skin to separate it, opening the skin as much as possible around the incision site. Throughout the skinning/ peeling process, you can carefully use your blade to cut through the connective tissue if needed, or use borax as an abrasive to help get the skin to come off the carcass easier. To do this, sprinkle some of the powder where you see the excess moisture and gently rub it in. You should see that it absorbs the moisture and that rubbing it against the carcass and under the skin helps the flesh separate easier.

3. Grab the ankle of one of the rear legs from the outside and use it to push the limb up into the carcass, towards the incision. You'll likely see the knee come through. Use your finger to work around the knee joint, separating the skin from the carcass up until the ankle. Once at the ankle, you can use a small pair of scissors or a blade to cut through the joint. Sometimes you'll be able to dislocate this joint, but the bones are small and soft enough to the cut through and around with scissors. No matter the tool, this leaves the foot attached to the skin. On anything larger, or on competition work, we would skin all the way to the toes but for commercial grade taxidermy, and for your first mount, leaving the small amount of muscle, tissue, and bone in the foot is totally fine! It will dry out as it is mostly cartilage, bone, and not a lot of meaty tissue. You'll repeat this step on the other leg.

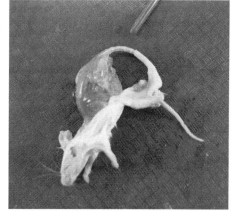

4. Now onto the tail and genitalia. You'll use your fingers to peel the skin away, and when you see the urinary tract, rectum as it connects to the anus, etc. carefully cut those "tubes" where they meet the skin. Male genitalia will require a little

more maneuvering, as there is more of it. Once the genitals are disconnected, we will skin the tail. Imagine removing a long sock or stocking without turning it inside out—that is what we will do. Use your thumb and index fingers to work the skin down the tail, exposing the bones that connect to make the tail. When you are about 1/3 of the way down, you will hold the skin firmly around this "tube" and gently pull the meat, using the edge of your thumb or nail to keep the "skin tube" right side out. Your index finger is for support and your thumb controls the pressure. It will take a bit of time and practice, and confidence, but the tail meat should easily slide out of the skin.

5. Now that the bottom of the body is free, we will peel the skin away from the belly and work up towards the front limbs. The skin and carcass are thinner here, so take your time and don't apply too much pressure at once. Continue to slide your fingers between the skin and carcass, and peel away rather than pulling.

6. The front legs can be a little tricky since they are much smaller, but the idea is the same as the back legs. Hold the feet from the outside at the wrist, and push the limb into the carcass. You'll see the elbow come through; you can grip that then continue to separate the skin down to the wrist. Once at the wrist, you will use small scissors or your blade to cut and disconnect it. Unlike the larger ankle bone, the small bones here are harder to dislocate so you may end up cutting though bone as you disconnect the limb. Again, this leaves the paws attached to the skin. You'll then continue to peel the skin up the chest, neck, and to the head.

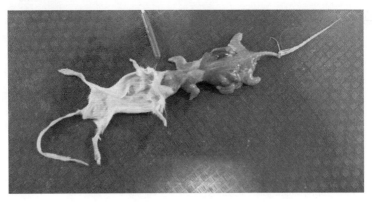

7. Be very careful with the head so that the facial features do not get damaged. Work slowly and gently, and be observant of the anatomy. If you get lost you can always re-invert the skin back over the carcass and use your fingers to gauge where you are. As you peel the skin up, keep the top of the skull facing you. You will see two areas on either side of the head where the skin is attached to the skull; these are the ear butts. Use the tip

of your blade to cut them a bit farther back, leaving some tissue attached, and you will then see the ear canal.

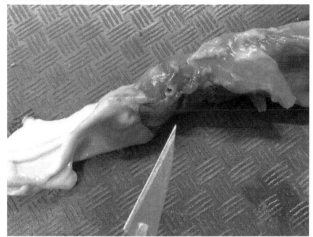

Once both ears are released, work the skin down to the eyes, and gently peel away. Some people use their thumbnail and forefinger to do this; others use their scalpel to cut though the connective tissue. You may end up cutting through the eyeball, but that does not matter as we want to keep the eyelids intact and undamaged. Be sure to stay close to the skull in order to keep the tear duct with the skin and not create a hole.

The next area you will come to is the mouth. Cut through the cheek meat at the corners of the mouth, and as the inner cheek and lip move away, you'll see the tongue and teeth. Work around the gumline, keeping as much of it attached to the skin as possible, and you'll soon be attached by just the nose.

Use your blade to cut in towards the skull to keep the nose cartilage attached to the skin. We will trim and clean this later, but while skinning, you want to use this bulk to prevent any damage to loss of the nose. The skin should come right off, or you can use your thumb and forefinger to gently pinch it off.

8. Congrats on skinning your first mouse! You may now put the skin, inside out except for the tail, into the alcohol to toughen up and help the fleshing go easier. Let it sit in alcohol for 15 minutes, or up to overnight depending on your time constraints. If you soak it overnight and realize you can't continue, wash the hide in Dawn soap and water, then freeze it in a zip bag or other freezer safe container until you are ready to work on it again. Soaking a hide this small in alcohol for too long will cause excessive shrinkage, and toughen the skin too much. You can bag the carcass and put it in the freezer so it firms up--it'll serve as good reference for sculpting later on.

Fleshing

Fleshing is a crucial step in taxidermy as you are removing any fat, connective tissue, and residual meat off of the skin. You are also making the skin more of an even thickness, which is important for the efficacy and penetration of a tan, and so that the skin dries evenly too.

1. Hold the skin taut over a beam or surface, and use your fingers/fingernails to peel away fat and connective tissue. You beam can be a wooden egg, an plastic horn, a piece of PVC pipe, the rounded end of a baseball bat, or anything you have around that you can stretch the skin onto so you have leverage while scraping it. You can also use a dull curved blade, like a small  butter knife or dulled scalpel, for fleshing, but on mice that is usually not necessary since their skin is so thin, and the fat is minimal. You can also use borax for abrasion, as it helps things roll right off the skin.

2. Once you have fleshed the body, use curved cuticle scissors or a small tool to trim any excess tissue around the face focusing on the ear butts and nose. For the lips, you will use your fingernail or a dull blade to gently "unroll" the gum line tissue so that your lips have a bit of an extra border around them. These "lip flaps" will be tucked into your sculpted head to re-create your lip line. You may also notice the skin near the whiskers is a bit meaty-by gently brushing in one direction with a soft wire brush to gently thin this area out, you will see the whisker butts or follicles become more clean and visible.

3. Once fleshed, soak the skin in alcohol for 5-10 minutes to help further tighten the follicles and clean any blood off the fur. Then wash it in the sink under a running faucet of lukewarm/cool water and a dime sized dot of Dawn soap to help degrease and remove fat, stubborn blood stains, etc. As long as the water is not too hot, you can give the skin multiple washes until it is clean. However, you do not want to let the skin sit in water for too long, or rub too hard, as water is a bacteria friendly environment and can encourage fur slippage. Be sure to remove any soap residue thoroughly. After washing the skin, you can gently squeeze it in a towel (do not wring) and let it air dry as you prepare the mouse body form.

Sculpting Your Form

Your body form is going to be strong and lightweight replica of the carcass. It is a good idea to freeze your carcass into your desired pose, and sculpt your form using that as reference. Looking at reference photos online helps too, as do anatomy books and illustrations. Here we have made our form in multiple parts using different materials.

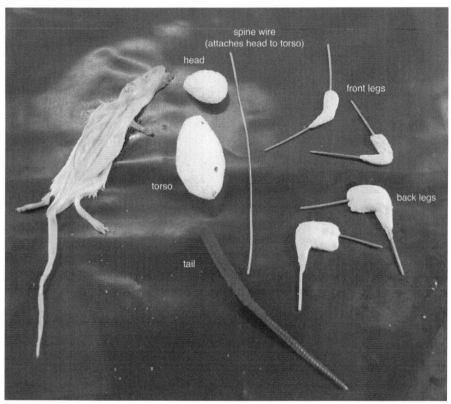

The head and torso are carved from carving foam (which can be bought at craft stores or from online taxidermy suppliers). The head will eventually be covered in a thin layer of a low shrinkage air dry clay in order to sculpt the finer facial features after sewing (we do this right before the skin is mounted to the form, to prevent the clay from drying out too much). The torso is marked with where the head, limbs, and tail should be attached, and we have sanded it so that the surface of the foam is a little rough (you can use a tool called a stout rougher, but for something as small as a mouse, a rasp and 80 grit sandpaper are more than enough). The limbs are sculpted using Apoxie Sculpt around 20g wire (these should harden and cure completely before you put them into the skin). Note the excess wire at the wrist and ankle area--this will be used to help our skin get onto the form, and will help your

mouse stand on a base or hold things in its paws. The tail is a tapered pipe cleaner. All the body parts are attached using hooked wire. Since our foam forms are so small, it is advisable to pre-drill holes using an awl or sharp piece of wire in a twisting motion. This will prevent the form wire from damaging the foam too much, though any minor damage can be filled in with Apoxie sculpt.

Preservation

To preserve the skin you have a few options. You can tan the skin using a commercial chemical concoction from a supply company, dry preserve it with a powder mix from a supplier, or use borax after an alcohol soak. Here we will use the alcohol and borax method. After washing and drying as detailed in the last step of the "Fleshing" section, powder the skin with clean borax. It is convenient to store your borax in a container with a tight lid, as you can put the skin in this container, close the lid, and shake it vigorously. This will ensure the powder reaches all areas of the skin, and you will notice how fluffy the fur is beginning to get! A light coating of borax is all you need on the underside of the skin, as you want to make sure it is still fairly flexible for assembly and mounting. (For this reason, it's advisable to make your form and powder it right before mounting). If you will be test fitting your skin onto the form, do not apply the

borax until you are absolutely certain everything fits (Borax dries the skin out pretty quickly, and you want to make sure it is flexible while you sew.)

Assembly

Once the pelt is preserved and our form is made, you are ready to mount, or "taxi" the "dermy". As our skin has been processed and has no memory, it takes a bit of patience as we are guiding it where it needs to go, where we want it to fold, and paying attention to the fur patterns.

If you are unsure of the accuracy of your form, you can "test fit" the skin onto the form. To do this, you will follow these assembly steps omitting any glue. If you choose to preserve your skin using borax or dry preserve, do your test fitting before applying the powder (as the clock starts on drying once the borax or powder

is applied). If the form does not fit, make a note of where it looks too large or too small, and adjust the form by carving, sanding, and/or sculpting.

First, install the tail by carefully sliding it into the empty skin, using a small amount of latex caulk at the tip. The caulk will help lubricate the tail form for insertion, and fill in the very small space at the tip of the tail so it doesn't shrivel. You can now move onto mounting the skin on the form.

On something small like a mouse, it is helpful to have the head skin inside out, and use your finger to help slide it over the head form. Roughly align the facial features, keeping the lip flaps visible. These will be tucked in later.

You'll now work each paw down onto the form, going from front to back. Hold the front paw between your thumb and forefinger, and make a gentle twisting motion to guide the wire through the paw pad or "palm". Slide the skin up and around the front legs, making sure the fit is snug on both front paws and limbs, and note where the skin is

<----"lip flap" that will be tucked in later

creased and tucked in for the underarm area.

Align the skin of the belly and torso, and use a similar grip and gentle twist motion as used on the front paws to insert the hind limbs and paws. The wires here will come out through the bottom of the heel (when you become more advanced, you can have the wires come out through the

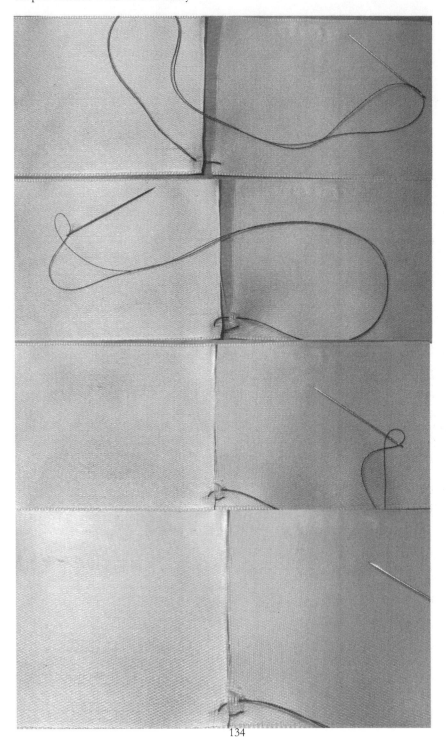

toes). Be mindful of where skin is creased and tucked in for the groin area. Once the hind limbs are on, insert the tail at your torso attachment point. You can use a small amount of caulk here to help secure and strengthen the attachment of tail to the form.

Once the skin is on and everything is close to where it belongs, you can start sewing up your incision. For this mouse we are using a doubled up thin nylon/poly blend thread and a basic sewing needle. Using too thick or too thin of a thread or needle can produce undesirable results. The baseball stitch, shown on fabric on the previous page, is the easiest and most commonly used. Insert the needle through the flesh side of the skin, pulling it up and through, then passing it to the other side. The needle always passes from the flesh side to the fur side, and in order not to tear the skin (as the thread can tear through a delicate mouse skin) you will pinch the skin together and then gently tighten the stitches. If you pull on the thread without supporting the skin, it can tear though the skin. Take your time and space your stitches about 2-3mm apart, keeping the fur out of the stitches. Most people prefer to sew from the head end of the incision to the tail end of the incision as this encourages the fur and hide to lay more naturally with gravity, and because the finishing knots can be better hidden by the butt. To finish with a knot, take a stitch but before it is complete, pass the needle through the "loop". Do this 2-3 times to fully secure the knot.

You can now use the wires to stand your mouse on a piece of wood or foam (the wires will pass through holes poked or drilled onto the foam or wood). We will proceed onto detail work!

Detail Work

The facial features will be set by using modeling tool or even a toothpick. For our mouse face, we will trim the "lip flaps" so they're about 2mm all around and tuck them into the clay. If your face does not look full, you can also add a small amount of clay into the nose, cheeks, and lips through the mouth opening from the outside. The tucked skin is secured into place with an insect pin. The pin passes ONLY through the tucked skin so that when it dries, no holes are visible in the lip line.

We have chosen to set the eyes from the outside, and are using black glass pin heads. Depending on the size of your mouse, you can use black glass beads, or custom made bubble eyes if you are doing competition work. We have trimmed the pin, and are using it to guide the eye into the opened hole between the eyelids. Use a tool to carefully pull the eyelids over the eyes, and shape your desired expression.

For the ears tuck the earbutts down into the clay. To keep the ears perky, we will support them by sandwiching them with wire mesh or painters tape, which can be curved to create the desired ear shape. Once you are more advanced, you can split the ears and make small earliners, but carded ears on a specimen as small as a mouse are acceptable for most applications.

For the body, be sure the armpits and groin areas are tucked into place, and that the fur patterns lay smoothly. Depending on the level of detail you want, you can even use sculpting clay and carve the anus so it can get tucked and shaped in a similar manner to the mouth.

Once all body parts are in place, the fur can be groomed with a blow dryer, or better yet, a soft toothbrush and borax that is "back brushed" in the opposite direction of growth. You can groom as the mouse dries so that the follicles are locked into place as everything dries.

To dry your mount, a cool, dry place is best. It can take anywhere from one to three weeks for your mouse to fully dry depending on the preservation method used and environmental factors. You will know it is dry once the paws are rigid. As it dries you want to make sure you monitor any shifting parts and take a few minutes each day to make adjustments. The hide is an organic material, and without our diligent monitoring, it can move unexpectedly while drying.

Finishing

After drying, you will notice the inner ears, nose, and bare parts of skin like the paws have faded or changed color--this is totally normal (since the preservative has taken effect, dried the tissue, and there is no longer life and blood flowing though the mouse). If you have an airbrush, thin washes of paint in these areas will help you build up a realistic color. Without an airbrush, Pan Pastels or powdered pigment (even powdered makeup such as pinkish blush and eyeshadow!) are a great

way to get natural, buildable, bendable color. Study your reference photos to get the appropriate color. Aside from color, you can also rebuild any shrunken parts of the eyelid or lip line with Apoxie Sculpt, and clean any clay on the eyes. A clear gloss coat makes the eyes shine, and can help protect them from dust. Finishing your mouse once it is dry will really bring the small details to life! After finishing, your taxidermy can be placed into a custom made habitat, such as a natural forest floor, or anthropomorphized with dollhouse props and accessories.

Care and Storage

We don't all live in museums, but there are a few simple steps we can take to ensure our hard earned handiwork is long lasting! Taxidermy is best stored in a cool, dry place, away from direct heat and sunlight, or drastic changes in temperature and humidity. Clean your taxidermy regularly with soft brushes dusted with borax, or a blow dryer. You can also use specialized products, but just using borax or a blow dryer regularly can help ward off moths and carpet beetles that like to snack on animal fibers. Pheromone baited moth traps are also great to help monitor for possible infestations, and since they trap male moths they do offer some protection against the larva that does the damage. For small animals, displaying them in a case or cabinet is ideal since it offers protection from dust and pests, and a mothball or moth crystals hidden in a cased habitat can offer an added layer of protection (do not use moth balls out in the open as they are extremely toxic!).

Congrats on your first mount! Taxidermy is a very rewarding craft, and you'll find that the more you do it, the more you'll learn. It can be a hobby to build personal collection, start a path to build the skills and confidence to attend and compete in your local taxidermy competitions and start a small business, or something totally unique to you! No matter what, I hope you'll use each mount you create as an opportunity for education and reflection on the beauty of nature.

Wet Specimen Preservation and Diaphonization
By Escher Null

Wet specimens are a great way to start off your collection, as well as to hone your skills as a taxidermist and preservationist. Eye-catching, simple to create, and wonderful conversation starters, they aren't what people typically think of first when you mention taxidermy and can help open people's minds towards the practice and art of preservation.

There are many different things that can be made into wet specimens. The most popular are fetal animals, organs, dissected examples and both juvenile and adult full specimens, but many sorts of animals, both vertebrates and invertebrates can be made into wet specimens. My main focus in this essay will be upon invertebrate specimens, though I will briefly go over preparation for vertebrates vs. invertebrates, as the process differs. I will also introduce the process of diaphonization, which tints and exposes the vertebrate skeleton. By the end of this, you will have the knowledge to begin your own projects in preservation of both types.

As for all pieces, I highly recommend you source your specimens from a reputable, humane, sustainable supplier. Veterinarian offices may have unclaimed animals or strays, but keep in mind that if you go that route, you will not be able to sell any of these pieces and they will have to remain in your personal collection, or be donated to an educational institution. Most US laws state that animals that were previously pets cannot have their remains sold. Reptile feeder supplies are wonderful, especially when first beginning wet preservation. You can purchase frozen mice, rats, rabbits, and other animals that have been humanely euthanized. Animal s that passed away at pet stores are also a viable option. Sanctuaries will occasionally have natural deaths, particularly bird sanctuaries, but be mindful of the laws in your country or state surrounding particular species of birds. If you do go to a pet store or sanctuary for remains, remember to state clearly that you are a preservationist, not just any old hobbyist.

If you lack the resources, fresh roadkill (no visible decay) is a viable source, but please check for local roadkill pick up laws, and whether you need to purchase a roadkill or salvage license. For invertebrates, I recommend going to a fish market, Asian grocery, or even an aquarium supply shop that may have recently deceased animals. The most important thing when it comes to your specimen is to make sure the subject was frozen as soon as possible after death. That way you can avoid decay and possible bacterial contamination. If your specimen has begun to smell strongly, has larvae or maggots, or is openly weeping fluids, it will not preserve correctly and you will be left with a jar of foul-smelling, disease attracting soup. As for the size of your specimen, your only limit is really container size. The larger a specimen, however, the longer it will take to preserve it. I've seen displays as large as a saltwater aquarium! For the purposes of this piece, I'll be focusing on smaller beasties, ones that can fit inside a pickling jar.

A list of things you will want to acquire:

- Specimen! Be it an eyeball from a large animal (cow or sheep are most easily acquired), a frog, fetal pig, or small furry critter.
- Jars! Your containers should seal completely. Use glass only! Tinted glass or UV-protective glass are your best bets, as these will limit sun exposure. Sun can damage your specimen and heat your fixative.
- Respirator mask. Inhaling chemicals is never a good thing.
- Nonporous nitrile gloves
- Plastic-backed puppy training pads, for fluid absorption.
- Syringe with hypodermic needle, gauge 18-26. Use a smaller gauge needle if the specimen has delicate skin that will easily show puncture marks. Ebay is a good source for these.
- Your fixative of choice. Formalin or ethanol work best. Ethanol may be easier to acquire, and both will do the job well. You may have a well-stocked fish store nearby, but it is likely that you will have to purchase formalin online. A buffered 10% solution will suffice in most situations. If using ethanol, you can find it at hardware stores. Try to find 70% diluted ethanol, and if it does not come diluted, ONLY dilute with distilled water. It should be noted that, if you are under the age of 18 (or 21, please check local laws) you may be unable to purchase these chemicals by yourself, and will need a parent/guardian/etc. to purchase them for you.
- Distilled water

You're going to want to thaw your animal until it is cold but pliable—any warmer and you risk bacteria all over again. Thawing in a sealed plastic bag in a cool place will work just fine--typically I seal mine and place it on the kitchen counter, or in the sink in case of leakage. The fridge will also suffice, but it will take a little bit longer. As a rule of thumb, don't thaw the specimen touching water, as this might cause premature decay. Periodically check on the firmness of the specimen: if it's still moderately firm and cool, not cold or warm to the touch, then it's perfectly thawed and ready.

Try to work on your chemicals outside, where ventilation is best. If that isn't available to you, open as many windows as possible and utilize fans to keep air circulation going. You'll need to inject the specimen with the fixative in every orifice and all major muscles. If the eyes are to be closed, also inject the eyes. Inject through the ears, mouth, and anus as well. The specimen should be very full of fixative, to the point of looking swollen. In the case of an individual body part or organ, simply inject in any open parts of flesh; for instance, a single limb, paw or tail would be injected in the cut end and possibly into the pads, in the case of a paw. An organ such as a heart will typically not require injection, and the same goes for invertebrates. Simply soak these in your fixative.

Next, find the jar you want to display the specimen in. Please use glass for this step, as plastic is unlikely to remain solid and nonreactive to your fixative. Both formalin and ethanol are likely to cause the plastic to leech into your fluids and specimen, if not melt the container completely! You'll need to pose the animal in the position you want it to remain in, as formalin will cause all the tissues of the body to harden into place. An easy pose to start with is to simply have the specimen curled up, as if sleeping. Look at photos of live animals to get inspiration!

Cover the placed specimen in more of the same sort of fixative you injected it with, and then it's time to seal it and wait! Seal your jar by screwing the lid on, and cover the lid with a washcloth and secure it with a rubber band, in case of any spilling. Store your specimen in a cool place, away from direct sunlight or heat sources. Gently shake the jar every other day to ensure fluid exchange and agitate your fixative. Depending on the size of your creature, this will take days to over a month. You will know it's ready when the solution is slightly cloudy.

Next, it's time to transfer the liquids! Your fluids have become cloudy with tissue and, well, other fluids. You can either filter the current solution, or use all new fluids. Filtering the solution is simple. Take a coffee filter and place it over a bowl or other jar, and pour the solution through the coffee filter. If you want to switch containers to a different display jar, this is the time. Fill the container you are now using with isopropyl alcohol. Typically I use a 70% solution. After roughly a week, agitating as before, you can swap out the fluids again for formalin, and begin their final soak. Don't be alarmed if you notice blood or clouding of the fluids, as this is normal. This time, do not reuse your solution, and use all-new fixative to preserve the clarity of your specimen.

Invertebrate Preparation

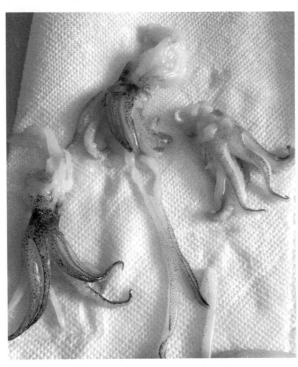

Now, after covering basic preparation, I will introduce the process in which I preserve invertebrates, in this case an octopus. Octopus can be an ideal specimen to start with, as they are readily available frozen no matter your location. If you cannot find any at a local grocery store (Asian grocery stores in particular often have them), try calling in and seeing if they can get one in for you. You may order them online, again from Asian groceries, but shipping a frozen specimen is pricey. Individual tentacles are a great starting point for invertebrate specimens, as they are inexpensive to purchase from your local fish market or Asian grocery. You may place the tentacle (thawed, but still cool to the touch) into a small jar and pour your formalin or ethanol mixture over it, replacing fluids as it grows cloudy. That's all!

However, you can also preserve a whole octopus. Depending upon the color that you want your specimen, you may want to gently set it in water before you add your formative, keeping the water just below a boil, to allow the skin to turn the vibrant deep red many people find the most attractive. I prefer mine to have their original, grey-blue tint, and for the purpose of this essay will be treating the process as if I have not boiled them. Boiling them gently doesn't affect their makeup or longevity, but avoid overboiling, which may make the specimen soft and mushy. This preparation is specifically for octopi, and doesn't include other sea creatures.

I was fortunate enough to have some very sturdy glass pickling jars to keep it in, and was able to properly direct the tentacles of my beastie to curl in a very beguiling way, using long cooking chopsticks to gently prod the tentacles into the configuration I wanted. It is not entirely necessary to inject your octopus or other invertebrate specimen with your formative, though as mentioned above, this is the typical wet specimen preparation.

An octopus is all muscle, and becomes incredibly rubbery as soon as you add your formative. My piece became VERY rigid and rubbery after that first week, which is normal and should be expected. Replaced the fluids gently, nudging the limbs back into their original formation with a long crochet hook or similar tool, and pour a mixture of 70% ethanol alcohol and distilled water, enough to cover your specimen, into the pickling jar. Place a label on the jar somewhere stating when the last liquids change was done, for easier record keeping.

As an aside, for the care of arthropod specimens, I would advise only using ethanol or isopropyl alcohol to preserve them, as formalin will stiffen them too much and may cause them to become fragile and hard to care for. Soft- bodied insects (aphids, thrips, small flies, and mites) become stiff and distorted if preserved in 95% alcohol and should be preserved in alcohol of a lower concentration. Matured moths, butterflies, mosquitoes, moth flies, and other groups with scales and long, fine hairs on the wings or body may be worthless if collected in alcohol regardless of the concentration. Keep these in dry preservation.

Diaphonization

Moving onto the next topic, we come to one of the most visually arresting types of preservation: diaphonization! Developed by scientists G. Dingerkus and L.D. Uhler[20], the animals are rendered clear by soaking in trypsin, a digestive enzyme. They also soak in stains of alizarin red and alcian blue, as well as other lesser-used

---

[20]   See: https://www.atlasobscura.com/articles/dyeing-the-dead-the-artful-science-of-diaphonization

dyes. They can be a bit tricky, but are a wonderful way to learn other methods of wet preservation. For this preparation type, I would recommend mice, rats, and frogs as good beginner pieces!

Items you will need for diaphonization:

- Exacto knife/scalpel to skin and gut your specimen. Many can be bought on eBay or through Amazon, so it becomes a matter of preference.
- Distilled water
- A solution of 10% Formalin
- Ethanol (95%)
- Alcian Blue 8GX
- Alizarin red
- Glacial acetic acid
- Measuring cylinder
- Glass jar
- Forceps
- Small kitchen scale
- Weighing paper
- Borax
- Trypsin, a digestive agent. (You can purchase it on Amazon in small capsules.)
- Potassium hydroxide, or KOH, purchasable on Amazon
- Thymol crystals, available, again, on Amazon
- Glycerin
- Hydrogen peroxide, 3%

The more complicated chemicals listed can be purchased on Amazon, eBay, and of course, your local chemist or pharmacist, if you have the proper licensing to purchase from them. Depending on your area, you may also find these items (trypsin, thymol, Alcian blue/acetic acid) at a hardware store or aquarium supply shop.

The beginning steps of diaphonization are much like your typical wet preservation; however, you will need to skin and gut your specimen before submerging in the chemical solution. Take care to not remove any of the bones, especially small ones such as those in the tail or toes, then rinse the specimen in water. Do your best to remove fat from it as it will impede the process of turning the remaining flesh clear.

After this, place the specimen in your jar with your 10% formalin solution, close the lid, label your jar, and let sit. Since diaphonization is typically implemented for smaller subjects, it should only take two or three days of soaking. After fixation the specimen needs to be washed in order to re-hydrate the tissues and to remove excess formalin. Remove the fixative from your jar and replace it with distilled water, gently rotating the jar to agitate the fluids and leaving it to sit for a day. Come back to it, rinse the specimen again, and replace the water for another day.

The next step, and the fun part, is the first stain! You will need a solution of ethanol (95%), Alcian blue 8GX, glacial acetic acid, a measuring cylinder, your jar, forceps, a scale, and weighing paper. A lot of stuff for such a small beastie!

Start by measuring out the Alcian blue powder, using a scale precise enough to measure 0.01 grams. Measure out 120ml 95% ethanol and pour into the jar with the Alcian blue, then add 80 ml glacial acetic acid and mix together. Transfer the specimen over this dye mix, making sure that it's completely covered and let it sit for one day. Your dye solution can be used more than once, but discard if you notice an excess of cloudy tissue buildup in your jar.

Now, a series of baths over four days! Your first soak should be in 95% ethanol, overnight. The next day, change out the ethanol and let soak another night. Day three, add 50 ml distilled water to your jar, pouring a little ethanol out if there is no room and let sit overnight again. On the final day, pour out all the ethanol and fill your jar with only distilled water.

Next, you will begin the protein digestive stage, using trypsin. Trypsin is a digestive enzyme that breaks down proteins, but leaves collagen. The breaking down of muscle and other tissues will help to make the animal transparent, while the collagen will keep the specimen from falling apart. You will need to make a

200ml mixture of borax (60 ml), distilled water (140 ml), and trypsin (2 g). The borax is added as a pH buffer for the trypsin, since it typically performs better at an elevated pH. You may use pH strips to test this, and the ideal range is ~7.5-8.5. Make sure you use borax and not boric acid, as these are two different chemicals.

Mix your borax, water and trypsin in an empty jar, then transfer your specimen to the mixture. Sit time will depend on the size of your specimen, usually not exceeding three days. Again, your mixture will be a little cloudier than usual. After those three days, it's time for bone staining with your alizarin red. You will need potassium hydroxide (KOH, 1.5 g), alizarin red S4 0.025 g), and distilled water (250 ml), in a 250ml mixture. Add these powders to a jar full of distilled water and shake gently. Transfer your specimen to the mixture and let sit for one day.

The specimen will go through a series of baths to make it transparent; this process is known as clearing. The longer the time spent in these baths, the better the clearing – please exercise caution, so as to not to leave specimens for too long as this might make them come apart. Check it-- does it seem to have soft bits of tissue floating, but still attached? Do the limbs seem loose, and do they move a fair amount when you move your container?

First, you will need to make KOH solution:

- 0.5 KOH (1.5 g)
- Distilled water (250 ml)

Bath #1:

- 150 ml of your KOH solution
- 50ml glycerin
- 2 ml of 3% hydrogen peroxide.

Incubate your specimen in this mixture for one day to a week at room temperature.

Bath #2:

- 75 ml KOH solution
- 75ml glycerin
- 1.5 ml peroxide

Soak again for one day to one week at room temperature.

Bath #3

- 50 ml KOH solution
- 150 ml glycerin.

Soak again up to a week.

You're at the final stretch! Take your specimen out of the jar for inspection, snipping away any excess tissue and imperfections. Once you're happy with your piece, place it in its final jar, filling it with glycerin and a small pinch of thymol crystals, which inhibit fungal growth. Use your forceps to play with the placement of your beastie, and fret not if they're a bit on the buoyant side. There will be small air bubbles in cavities that will dissipate over time. Wait a few days for the bubbles to clear, then reopen your jar and reposition the specimen.

I hope you have found my brief guide to wet preservation and diaphonization helpful and that is has helped you refine your methods, and that it enables you to practice your preservation more easily in the future!

Dry Preserving Wings
By Lupa (yes, we're back to me now!)

Dry preserving bird wings is so simple and easy that even I in my eternally busy state can do it! Obviously you want to make sure that the bird you are working with is both legal to possess and was procured in a legal manner; if you're in the United States, please leave that roadkill hawk or owl where it is! I've primarily preserved chicken wings, but this process will work for almost any bird wing. Here's what you'll need:

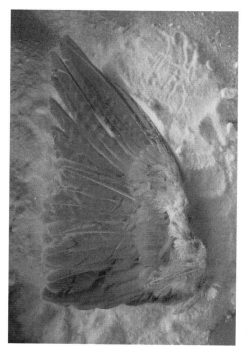

- A cardboard box whose bottom is a little larger than the outstretched wing
- Borax (the laundry stuff, not boric acid from the pharmacy) – some people use table salt instead
- T-pins
- One large piece of styrofoam or several decent-sized ones (optional)

Start with a whole wing that has been cut from the bird and cleaned of blood and debris. The fresher, the better is the rule here. Once the flesh on the bones starts to rot the smell will seep into the feathers and you're probably never going to get rid of it. If you can't preserve the wing right away, stick it in an airtight bag in the freezer. You may wish to do so anyway to kill off parasites. Put it in for two weeks, take it out for a week so any eggs may be prompted to hatch, and then freeze for another two weeks. That should kill just about anything; whatever's left won't survive what's next.

Put an inch of borax in the bottom of the box; you can make the layer a bit deeper for a really big, muscular wing like domestic goose or turkey. Break up big clumps of borax as best as you can. Lay the wing on top of the borax and spread it open however far you want (unless you want to preserve it folded.) If you have trouble getting it to stay open, put your styrofoam in the bottom of the box, then

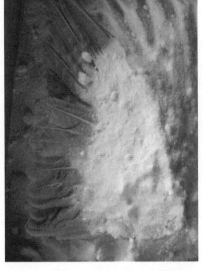

pin the wing to those using T-pins; you may need a thicker layer of borax to cover the wing as it's lifted up on the styrofoam. For smaller wings you may just be able to stick the pins through the cardboard and have them stay; just be careful what's underneath the box doesn't get stuck or scratched.

Now put another layer of borax over the wing, at least an inch thick. You want to make sure in particular that the bony, fleshy part of the wing is covered; covering the rest of the feathers entirely isn't quite as crucial. Close or cover the box and put it in a cool, dry place. Check it every two to three weeks; when the wing is stiff and can no longer be folded or bent easily it's ready to go. Bonus: the borax helps to desiccate germs!

Brush the extra borax off the wing; keep it stored in a dry place. Getting it wet can cause the flesh to rot and, of course, get smelly, so please don't try soaking to reshape it. Once it's dried, that's the pose you're stuck with whether you like it or not. The borax can be reused many times; just break up any clumps and let it air dry, or put it on a baking sheet in the oven at a low temperature for a half hour or so to dry it out.

As awesome as it is, this book is not the be-all and end-all of how to prepare hides, bones and other specimens. There's plenty more information on tanning, bones cleaning and the like, and we haven't even covered all the myriad ways you can preserve the remains of a dead animal. A few others you may wish to look into more are:

- Mummification: Most of the mummified animals you'll find floating around the sales pages were naturally preserved, usually animals that were unfortunate to get caught inside walls or water heater insulation and died. There are a few Vultures who have figured out methods of mummifying animals, but they tend to be pretty guarded about their methods. There are a couple of tutorials on how to mummify chickens Egyptian style for classroom projects, but I have no idea how well those hold up long-term.

- Freeze-drying: Taxidermists have used freeze drying for quite some time to preserve delicate portions of an animal's body like a turkey's head and neck. As with mummification you may need to dig through the internet a while and use specific search strings like "freeze-dried taxidermy". You'll also likely have to sift through sites for people who want you to pay them to do that service for you, rather than teaching you how.

- Bug pinning: While the previous methods in this chapter have largely been for use on vertebrates, bug pinning is a relatively easy method of preserving insects and other terrestrial arthropods. A search for "how to pin bugs" brings up a lot of tutorials; I recommend those that go into a little more detail rather than just saying "stick the bug on a piece of foam with a pin and let it dry." Be aware that many sources will expect you to start with a freshly killed bug, which you may not agree with. If you find your own naturally dead insects, spiders and the like, they can generally be preserved simply by putting them in a small jar or vial to protect them from being crushed. If an invertebrate specimen dried in a position you don't care for, there are also online tutorials on how to rehydrate and reposition them, though they do become more brittle with age and so many not take to rehydration as well over time.

- Blown eggs: This is the process of making holes in the ends of bird eggs and then carefully removing the contents so that the shell can be preserved more or less whole. It's a great way to make egg specimens, but it can be a bit fussy, so practice on some chicken eggs from the store first.

Appendix I has a list of books that go into more detail on how to do your own tanning and taxidermy. The internet is also a wonderful source of information; YouTube and Vimeo videos are especially useful as you can watch the process step by step. Some websites and videos may be better than others; if you're researching tutorials on how to create taxidermy using frozen rats, read/view several before trying one out, and if there are comments read them to see what other people's feedback is like. Be prepared for some trial and error, and start with cheaper materials to practice on so that you don't ruin more expensive remains with a lack of expertise. And don't be afraid to ask questions in forums and groups, since many people have likely faced the same problems you have and can provide constructive feedback.

Also, I know I keep talking about "look for stuff online" without giving a ton of links. As I've mentioned elsewhere, the collection of links in this book has a limited shelf life before it becomes obsolete. So I recommend using your favorite search engine[21] to search for things like "skeleton articulation tutorial", "how to

---

21 If you don't yet have a favorite, I recommend https://www.ecosia.org/ - the company uses the ad revenue on its search results to plant trees!

mummify an animal", or "how do I tan a hide with soy lecithin?"

Not all of these are apartment-friendly, obviously. Your best bet if you live in a confined space, especially with other people, is to stick to small tanning projects (feeder rats as opposed to whitetail deer), various sorts of dried specimens that are buried in a substrate like Borax that absorbs or hides smells and hinders rot, and anything that can be processed relatively quickly like wet specimens. You may also wish to limit yourself to either very fresh carcasses or ones that were frozen shortly after death, as anything rotting gets really smelly really quickly.

A Brief Note on Edible Animals

Some Vultures have a strict ethic of "use every part of the animal", and that includes the meat. There are many cases where that's just not possible, either because you don't have the entire carcass, or because the meat has rotted past the point of safe consumption. Some animals simply don't taste as good as others; trying to make coyote meat palatable is more of a challenge than with chicken.

If you are a hunter or farmer and you kill your own animals, then you can pretty safely process the meat for consumption. The same goes for buying bone-in meat at the store—cut the meat from the bones, cook the meat and get the bones in the maceration bucket. If you decide you want to take bigger risks and you can legally collect roadkill, I recommend Miles Olsen's *The Compassionate Hunter's Guidebook*, as he talks about signs he personally uses to determine whether roadkill is safe to eat, and he does an excellent job of talking about how to use literally every single part of a deer.

However, because I care about you and I don't want you dying of botulism, I'm going to end this chapter by saying: **never, ever, ever eat any meat that you cannot 100% verify as safe for human consumption**. When in doubt, put it outside away from where pets can get to it and let it return to the earth. Or, if unsafe chemicals may be involved, follow the instructions that Shelby described above.

Chapter 5: Put Your Best (Rabbit's) Foot Forward: Displaying and Caring For Your Collection

Once you start bringing home hides and bones and stuff, you're going to have to figure out what to do with your dead menagerie. While it may be tempting to just put everything in a pile, sit in the middle and spend hours and hours oooohing and aaaahing over your new dead friends, realistically you need to be *slightly* more organized than that.

This chapter will present some potential solutions to the "Where the hell do I put everything?" problem. Obviously your mileage may vary according to where you live, how much space you have, who shares that space with you, and what you're trying to find space for.

Unlike baseball cards, you can't just stick your entire collection inside a few boxes and tuck them in the top shelf of your closet. Not only would they outgrow the closet space fairly quickly, but some of them do require periodic maintenance. Let's start with some ideas on how to keep your collection from overrunning everything.

Public Display Versus Hidden Wonders

The first thing you want to think about is whether you want other people in your home to be able to see your hides and bones. Many Vultures don't particularly care, and scatter their collection all over the place. Some, particularly younger folks, may need to hide their dead things at least until they move away from parents and other housemates who may take offense. And other people find that their hides are best left in the closet lest their very much alive dogs or cats get hold of them and turn them into furfetti.

If you can't keep your collection at home, think about where else you might keep it. Is your workplace more amenable to dead things, especially if you have your own office? You may also see if any of your friends are willing to temporarily hang onto at least some of what you have until you can get into a better living situation. Unfortunately, it may be the case that you simply aren't able to practice your own brand of Vulture Culture right now; better to wait than to have your collection thrown out.

Just a Part of the Décor

So let's say you are able to have your collection out in the open. Some people like to scatter their hides, bones and the like around the home. You can do so in a creative manner, maybe a floor-standing taxidermy mount near the front door to greet visitors when they first come in, or a selection of skulls hanging in the hallway along with family photos. Just be mindful of where you place certain items; the bathroom is generally a bad idea for most animal remains as the damp can cause mold, and direct sunlight is bad for pretty much everything from taxidermy to wet specimens. Kitchens are also bad news, not just because of steam, splashes and other potential messes, but also because you don't want to contaminate your food with things like vintage arsenic-tinged taxidermy and alcohol and formaldehyde from wet specimens.

I know it's tempting to just dedicate your entire home to dead stuff, but avoid letting your collection take over. Make sure any larger pieces of taxidermy aren't blocking walkways and doors, and if you live with others check with them before turning any common areas into display space. Keep things off the floor as much as possible so that they don't become tripping hazards—and potentially damaged.

If you're really into interior decorating, you can be quite elaborate in your presentation of your collection. Glass domes and other tabletop display cases can make just about any smaller specimen look extra classy, especially if you add a bed of moss or artistically arranged dried flowers. You can also add skulls and bones to plant terrariums, though be aware the humidity may cause them to grow algae over time.

Of course, you can also just tuck bits of your collection in between the volumes on your bookshelf, or push the books all the way to the back and use the ledge in front for smaller items. Work with the space you have; you may surprise yourself at how many places in your home are just begging for a bit of hide or bone to complete them!

Cabinets of Curiosity

You might prefer to keep your entire collection in one dedicated place in your home, whether an entire room, or at least one corner in particular. This follows the grand tradition of the cabinet of curiosity, and makes quite a visual statement, especially with larger collections. (It also makes it easier to hide the collection if you have squeamish visitors or don't feel like explaining to the repair person why your home is full of dead animals.)

While you can certainly go all-out and have ornate hand-carved wooden display cases for your collection, there's nothing wrong with some secondhand Ikea shelves if that's all you've got. I've managed to find some pretty neat fixtures on Craig's List for not a whole lot. Some Vulture homeowners have even custom-built displays right into the wall, which may be an option if you're a DIY sort of person.

If you really want to stick to the theme, try organizing your collection according to type, or country of origin, or even individual species. Or use whatever organizational method strikes your fancy. You're likely going to have to balance that out with the realistic limitations of your space. As much as I'd like to have all of my animal skulls in one gigantic set of shelves, arranged by phylum, class, order and so forth, they're instead clustered on a few different bookcases and window sills, and not always with their closest relatives.

Because I like using my skulls in art and education, I've also added identification tags to each of them, and I maintain a database of what I have on my computer. This not only helps other people determine which skull is what species, but if my collection ever gets significantly larger it'll help me keep from getting duplicates!

Sacred Spaces

For some people, Vulture Culture is inherently spiritual. We see the hides, bones and other remains as sacred, and we wish to care for them as lovingly as we would the remains of our own families. If this sounds familiar, you can work your hides and bones into altars and shrines. As with other displays, be mindful of where you place them, especially if you don't want curious visitors picking everything up and looking at it.

It's best to make these altars indoors; most critter bits don't do well outside. You can have bones and antlers outside, but they will weather away over time. Of course, that may be your aim, in which case you can enjoy watching a patina of algae and moss spread over the increasingly worn bone as it returns to the earth.

The actual design of each altar is entirely up to you. Some people dedicate the space to a particular species, or a deity associated with animals. Others just put whatever feels right on the altar. It can be all hides, bones and the like, or a mix of sacred items. And it doesn't really matter whether you use a table, bookshelf, dresser, or even an entire wall—anything goes!

If you're interested in finding out more about the spiritual side of Vulture Culture, I wrote an entire book on it a few years ago, *Skin Spirits: Animal Parts in Spiritual and Magical Practice*. It is from a decidedly neopagan/animistic perspective, but isn't so much a book of dogma as a toolkit of ideas for you to incorporate into your own spiritual practice.[22]

Create Art

I'll admit this is my favorite solution for the problem of Too Many Dead Things. After all, I can more easily justify buying more if they're going to be art supplies! I have plenty scattered around my home and studio, and that's not including the ones I have for sale.

So what can you make out of dead things? Leather and fur both lend themselves well to sewing projects, from clothing to pouches to plush taxidermy. Bones make excellent canvases for painting, especially with acrylics, oils or earth pigments, as well as handles for rattles and other ritual items. Carving skulls and bones with a Dremel or other multitool has become popular in recent years, and you can find tutorials online. Feathers are a little more delicate, but I love them as

---

22 If you're interested in my other books, you can buy them directly from me at http://www.thegreenwolf.com/books, and I'll even sign them for you!

accents on larger projects; whole wings and tails make great fans. Some people also like to take legal feathers, like white domestic turkey feathers, and paint them to look like the feathers of birds that aren't legal to possess. Claws, teeth and smaller bones are excellent jewelry fodder, though you may need to invest in a Dremel or similar tool to drill holes in them.

I'll refer you again to my book *Skin Spirits*, as I include instructions on how to make several art projects with hides, bones and the like. You can also get kits with materials and instructions for simple projects from Tandy Leather.

Tips For Small Homes

Not everyone has a three thousand-square-foot home to show off their collection in. If you're in a tiny apartment and/or have to share space with a bunch of other people, you may find it challenging to find a place for everything. The single biggest tip I can give you is to make as much use of vertical space as possible as floor space will likely be at a premium. After all, you probably don't need to walk on your walls!

I find hides to be the toughest thing to display. I dislike hanging them by their noses, as I often reshape the faces on mine and hanging would just destroy my work. Wall-hanging quilt racks are a good option, and many come with a shelf perfect for displaying skulls and other items. A cheaper option is to get a broom handle or long dowel rod and attach cord loops or straps to the ends to hang on the wall. Then just drape the hides over the rod for a simple but effective display solution! I've also used the backboard of my bed as a place to hang smaller hides. And sometimes the corners of rooms end up with several folded-up pelts in a snuggly pile.

Skulls can also quickly take up a lot of shelf space. If you have some skulls without lower jaws, you can hang them on the wall with a bit of cord or leather strap. Lay the center of the strap on the underside of the skull, about level with the eyes. Then thread the ends through the holes created by the cheekbones. Tie the two ends together, and pull the strap snug against the underside where you started.

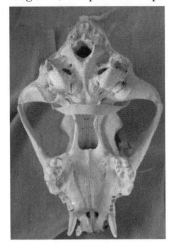 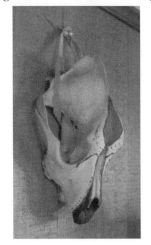

You can now easily hang the skull by a tack or nail.

Don't be afraid to stack skulls and other hard items. Just be very sure that they stack together securely, and that they're tucked back where people aren't likely to bump into them. You can give skulls a little more traction by layering a bit of leather in between them.

Little wall-hanging curio cabinets like some people use for showing off souvenir thimbles and the like are great for displaying small items like claws or teeth. I'm also fond of vintage printing type trays, and they're fairly easy to find at antique stores. Just string a wire or other hanger on the back, and you're good to go.

If you want to really save space, make use of your ceiling! Large hanging plant baskets can hold a pelt or two in an attractive manner. You can also tack or nail a pelt flat to the ceiling, kind of like an upside-down fur rug. If your ceiling is high enough, try hanging smaller, lighter skulls like ornaments from it. (Bonus points if you also string Christmas lights from the ceiling, winding around your display of suspended skulls.) Take a cord and tie each end to one of the cheekbones, then hang them from a plant hook; you can also tie the cord to a screw though make sure both the screw and knot are very secure. You may wish to use a stud finder to make sure whatever hardware you're using is set into wood and not just the plaster ceiling.

A Few Notes on Work Space

If you enjoy processing raw specimens, you're going to need a place to make a mess. Outside seems like it would be the most obvious choice if you have your own yard, but there are some challenges. The weather is the biggest obstacle most people will face. While a warm, sunny summer day is great for helping maceration buckets along and keeping you comfortable while you tan rabbit hides, the reality is that many days throughout the year will likely be less pleasant. It's not as much fun when the temperature is below freezing or absolutely sweltering and you're being attacked by mosquitos. Plan accordingly if you choose to work outdoors.

You may also have to deal with bigger (live) critters than mosquitos. Raccoons, possums, dogs and bears are just some of the animals known to try to get into maceration buckets or make off with raw hides. And if your yard isn't securely fenced neighborhood kids may come over and poke around in your stuff out of curiosity or boredom.

Speaking of neighbors, if the people around you can see you working with animal remains, they may be alarmed. Be prepared for everything from concerned questions to having the cops called on you.

If you're lucky enough to have a garage or shed that you can dedicate to your tanning or bone cleaning, that takes care of a lot of the above problems. But what if you have neither garage nor yard nor even a basement utility sink? Well, hope that anyone you live with is amenable to you using the bathroom as a work space. My partner has spent years with me hand-washing hides in the bathtub, and buckets of bones soaking by the toilet. Granted the hides were already tanned and just needed a quick bath, and the bones had a little coffin wax that needed to be removed, so I wasn't working with raw materials; that would have been a bit too much of a mess even for his great patience.

You may or may not be able to make use of indoor space for your work. If it just isn't going to work out right now, be patient, and wait until you can either move out or space can be made to accommodate you. As I said earlier in the book, you don't have to tan hides or clean bones in order to be a Vulture!

As a hide and bone artist, I've also had to come up with creative solutions for my art supplies, especially as I've had over to two decades to accumulate quite a stock. I'm fortunate in that I've primarily either lived alone, or with roommates who have been accommodating of my work and the need for a little more space. I do get a lot of mileage out of big plastic storage bins and milk crates; thankfully both can be had cheaply at local thrift stores. My workbench also has some shelving stacked on it for tools and other supplies.

One of the most valuable practices I've adopted is to label everything. There are few things more frustrating than having to crack open a dozen storage bins looking for one particular pelt or bone, only to have it be in the one in the very back at the bottom of the stack. I'm not quite as detail-oriented as to maintain a spreadsheet of current supplies, but you're certainly welcome to do so.

Care and Repairs

Time waits for no one, and that includes the already dead. While your specimens will already be preserved, they will need some upkeep now and then.

Generally speaking you want to keep your specimens out of direct sunlight, and away from heat or cold extremes. They'll appreciate being dusted now and then, and keep them away from pets who may chew on them (cats and dogs also sometimes like to pee on pelts!) Damp is also bad for fur, untreated leather, feathers and other "soft" items, and it can cause the keratin sheaths on claws or the horn sheaths on ungulate skulls to come loose.

Keep an eye out for clothes moths and carpet beetles. Both of these insects can destroy fur and feathers in an alarmingly short time. Clothes moths are small (about ½" long at the most) and hold their wings folded tightly to their body when they aren't flying; their larvae are tiny soft white wormlike things. Carpet beetles are even tinier, less than ¼", and are either mottled white, black and brown, or a solid black or brown, depending on species; the larvae are tiny hairy caterpillar-type insects.

Vacuuming hides with a hose attachment periodically can help keep them from infesting your collection. If you're concerned about moths in particular, you can store hides and feathers in a cedar chest or a box with cedar chips in it. You can also use mothballs, but they have more toxic components like 1-4 dischlorobenzene, camphor or naphthalene. If you do end up with an infestation, put any affected items in a plastic bag and then into the freezer for a week. Take it out again for a few days to let any remaining eggs hatch, then freeze again. Alternately, you can put them in a hot car on a sunny day for a few hours and the heat should kill any insects on them.

Here are some specific care tips for some of the more common animal remains:

Hides

Rawhide is pretty low-maintenance. It will expand and contract with heat, especially if it's stretched tight as on a drum, but this is normal. Just keep it out of extreme temperatures. You can both rawhide and tanned hides with any leather conditioner. There are plenty out there, like mink oil, saddle soap and neatsfoot oil. Some people even create their own custom blends. I like to oil hides in my collection about once a year unless they seem especially dry or greasy, but your timing may vary depending on how dry or humid your climate is, how old the hide is, the quality of the tan, etc. If the hide feels dry and scratchy, or if it's stiffer than usual if tanned, go ahead and give it a dose of conditioner. Most conditioners will have directions on the container. I prefer cream-style conditioners to liquid as it's easier to work the conditioner in evenly, but try both and see what you like better.

If rawhide or a tanned hide (leather or fur) gets a little dirty, you can just wipe it off with a damp cloth. For suede in particular, use a suede brush or other stiff-bristled brush to remove the dirt once it has dried. If it's a bigger mess, and the hide's tan is a newer one or still in good condition (the hide is stretchy and supple) you can probably give it a quick bath in water and a gentle soap like castile. Test a small piece of the hide first; soak it in water for about ten minutes, let it drip dry for a half hour, and then gently tug on it to see if it tears. If it doesn't, then carefully hand-wash the affected part of the hide.

Generally speaking, water is the enemy of tanned hides. If they do get wet, whether on accident or because they needed a bath, always air dry them, never heat dry them! It's not a bad idea to oil the hides afterward, either, using a good quality leather conditioner. This will darken leather hides; also, you do not need to oil the hair side of a fur, only the skin side. I like oiling hides once or twice a year anyway, depending on how dry the climate is, how often they're getting worn in hot summer weather, etc.

If you really want to go all out you can see if there is a local fur cleaner or storage business in your area who can give your hides an overhaul once a year. The same goes for taking leathers to a leather specialist. Contrary to popular opinion, I don't recommend having fur or leather dry-cleaned as it can strip out oils and otherwise damage the hides.

Although leather and fur can be quite strong and flexible, they can rip, especially garment-weight hides that are thinner. If this happens, you can sew up the tear using special needles and thread; I personally like John James brand glovers' needles, size 1, and artificial sinew that's been split down to its thinnest thread. Fur should always be sewn from the leather side, not the hair side. For more delicate hides that may not stand up to sewing, or hides that are too thick to sew, you can make a patch with thin leather and glue it onto the back side of a leather hide, or the leather side of fur. For this I really like tacky glues like Aileen's or Tandy Leather Weld.

Bones

Bones really don't need much care once they've been properly cleaned. Just avoid sun and too much heat or cold, and dogs that may decide they're excellent chew toys. Dust them every now and then, and if you like give really grungy skulls a bath in water and a gentle soap. Don't soak skulls for days at a time, as this can cause them to fall apart or lose teeth.

If a bone gets damaged or a tooth falls out, it's relatively simple to glue them back together, especially if it was a clean break. I generally use Gorilla Glue or another super glue, though some people swear by Elmer's, while others like industrial adhesives like E6000. B-72 acryloid is sometimes used in museums for its stability and reversibility if the pieces of bone need to be taken apart again. If there are bone chips missing, you can fill in the space with epoxy putty, and then lightly sand with a fine grain sandpaper once it's cured.

Occasionally you may end up with a skull that wasn't properly degreased. You'll know it by the yellowish staining that forms as the oils make their way to the surface. You can either send the skull to a professional bone cleaner for degreasing, or do it yourself using the tutorial in this book. The oils won't cause the bone to instantly crumble apart, but they're unsightly and can occasionally smell. They also can harbor bacteria that can damage bone over time.

There are also bones that have been painted or varnished. Acrylic paint can be removed by soaking the bones overnight in a 50-50 Pine Sol and water solution; the paint should peel off, though you can leave it in longer if the paint is being stubborn. I wouldn't leave it in for more than two or three days. For other kinds of paint and varnish, soaking it in acetone can help remove the stuff, though you may still have some scrubbing to do. Leave it in overnight, and see how the paint/varnish is, and leave it in longer if you need to. If you aren't sure about how a given skull or bone may react, just soak a tiny portion of it to test it. Make sure you rinse all the cleaners off once you've removed the paint.

Teeth are a special circumstance, since they can be a bit more delicate over time. They're hollow inside, and can develop hairline fractures along their length, especially if they're subjected to extreme temperatures or humidity changes. I've even received teeth that cracked while being shipped to me via air mail, since the baggage hold can be very cold. If this happens, carefully split the tooth in two pieces, using an Exacto blade or other very thin object (be VERY careful not to cut yourself!) If you have trouble getting it to split, you can cut off the very end of the root to expose the hollow inside, and use that space to pry the tooth apart. Now super glue the two halves back together, and carefully clamp them using a clothespin or binder clip, or just hold them together with your fingers (without super gluing the tooth to yourself, of course.) If there's still a visible crack you can fill it with a mix of super glue and baking soda, or white epoxy putty, and then sand it smooth once it's completely dried. To prevent cracking some people coat the tooth in acrylic sealer or resin. Because humidity can also play a part, some people oil the teeth in their collection with mineral oil.

Antlers and Horns

Antlers can be cared for in much the same way as other bones. Additionally, as with teeth, some people like to oil them to give them a nice shine and prevent cracking. Leaving antlers outdoors will cause them to fade and deteriorate, especially if they're out in the sun, though some people like this weathered look. Since deer shed their antlers every year, you may happen across some of these whitened, older specimens in the woods.

Horns are a different story. They're not shed, and instead are permanent keratin sheaths covering living bone. Most of the time when you get horns from bison, antelope, goats and other ungulates you're only getting that outer sheath. Keratin is the same stuff that makes up your hair and nails, and it takes well to conditioning. You can oil horns with mineral oil to keep them from drying out and fading. Be aware that certain insects also like to chew on horns, so inspect them periodically for damage. If you do find evidence of insect infestation, put the horn in a sealed plastic bag and then put that in the freezer for a couple of weeks.

If anyone ever offers to sell you a shed cow, buffalo or antelope horn (with the sole exception of a pronghorn), they're either misinformed or lying. Only antlered animals like deer shed their antlers every year. The pronghorn antelope is

the only horned mammal to shed the outer sheath yearly.

Also, for those of you who are particular about having shed antlers, a true shed will have a slightly rounded, pockmarked surface at the base, below the rosette (the part that flares out at the end opposite the tines). An antler that has been cut off a skull will be perfectly smooth there. The photo to the left shows a shed antler base.

Feathers

Feathers are also relatively low-maintenance once they're cleaned, especially if they're on display and not handled much. They need dusting now and then, and should be monitored for moth or beetle damage, but otherwise they'll be fine on their own.

Feathers that are worn or otherwise handled more often can get ruffled and dirty. To clean feathers, once again you want that water with gentle soap. Wash the feathers by hand, removing any visible dirt. Let it air dry. Then you can get the feather back into proper arrangement by running your fingers up the edge of the feather from the bottom (quill) to the tip. Wings can be cared for the same way, except do NOT immerse an entire dried wing in water; the dried flesh can begin to rot. Instead, clean off dirt with a damp rag, and carefully rearrange the feathers by hand.

Invertebrates

There are plenty of invertebrates in the world, many more than there are us vertebrates! I'm going to talk briefly about three different types of invertebrate specimens: wet specimens (which can also be vertebrates), pinned bugs, and shells.

- Wet specimens: the best thing you can do for your wet specimens is to not subject them to direct sunlight or temperature extremes, and keep the lids on tight. Keep them far, far away from flame and heaters as the chemicals used to preserve them are **highly flammable**. If the preservation fluid they are in becomes cloudy or discolored, all you have to do is pour out the old fluid and replace it with new (generally 70% isopropyl alcohol.) Safely dispose of the fluid by taking it to a hazardous waste disposal facility rather than simply dumping it down the drain, and definitely DO NOT pour it down a storm drain as those almost always empty directly into streams with fish and other living beings rather than the sewers.
- Pinned bugs: again, keep them out of direct sunlight to avoid discoloration and brittleness. You also need to keep them out of any particularly humid settings since they can get moldy. Many collectors keep their bugs in glass cases, both to protect them from humidity and also to keep them from being accidentally crushed, as they are extremely fragile. This also goes for shed tarantula skins and other exoskeletons.
- Shells: generally these don't need a lot of upkeep. Again, keep them out of sunlight and away from heat, and you can polish them with mineral oil if they get dull. Also be careful if you're carrying a bunch of them at once, such as in a basket or a box, as they may rub together and get dinged and otherwise damaged.

A Note on Insurance

As you will likely find out pretty quickly, it's easy to invest a lot of money in your collection. If you live under your family's roof their renter's or homeowner's insurance policy should cover damage and theft, though be aware of how much of a deductible you'll have to be responsible for. If you own your own home you should have provisions in your homeowner's insurance; if you rent, you should already have renter's insurance. If not, now's a great time to sign up! I pay less than $10/month for mine as an addendum to my auto insurance. Also, check with your auto insurance company to see what coverage there is for items stolen out of your vehicle.

It's a good idea to have at least a basic spreadsheet listing what you have in your collection. It's even better if you can provide receipts and prices for everything. That way if something big happens like a house fire or flood where everything is lost, you have a record you can add into your claim. You'll want to make sure you have an offsite copy of this record, whether on a thumb drive in a safe deposit box, or in cloud storage online.

If you are a working hide and bone artist, you'll also want to have small business insurance, especially if you sell your work at events or through shops. This protects your stock even when you've taken it out of your home. And in some cases it may even cover items sent to you for custom work that end up damaged or stolen.

People and Pet-Proofing Your Collection

Unless you live entirely alone, you are going to have to deal with placing some boundaries around your collection. In an ideal world everyone in the household will know the value of what you have and be respectful of it. Realistically you may have to put forth some extra effort to keep everything safe, especially if your home includes free-roaming pets, young and rambunctious kids, or adults who don't respect you and your stuff.

Sometimes the solution is as simple as keeping everything behind a closed door or up off the floor. If you have dogs, the goal should be to keep everything higher than they can jump. Cats are trickier, since they jump and climb and are pretty persistent. In that case, having everything in a display case or even a closet may be best, at least until you can train them to stay away from your collection. In this case, train them the same way you would train them to stay off the kitchen table or away from the garbage: spray them with water if they get too close, or tell them "No", or throw an old soda can with a few coins in it at the ground in front of them to make a scary noise.

Kids are trickier because they have hands with opposable thumbs and can do things like open doors and cabinets. Young kids are especially prone to this as they're busy exploring the world and don't always understand why they aren't supposed to do certain things. In addition to making your collection more physically inaccessible, you may need to teach the kids that they are only allowed to see your collection when they have your express permission. For example, maybe a couple times a month you can bring down a few things for them to look at and carefully handle. Tell them if they want to see anything else they have to ask you, and set a limit on how many times they can ask, like three items per week. Coordinate with other people in your home to make sure that these rules are enforced and that there are definite consequences for breaking them.

Needless to say, you may want to keep your rarest, most expensive and/or most sentimental pieces hidden away as best as you can until you either live somewhere else, or can thoroughly convince everyone in your household that your collection is off limits without your permission.

Traveling With Dead Things

Occasionally you may have a reason to bring something from your collection along with you on a trip, or when you're moving to a new place. Perhaps it's a piece of costuming for an event, or a gift for someone, or something you bought and want to bring home. Or you might be moving your entire collection to a new home. No matter the reason, there are a few things you'll want to keep in mind to keep your specimen safe and sound.

Legalities

If you are crossing state or international lines, you want to be very sure that what you have can be brought with you legally. Again, you can find a database of animal parts-related laws at http://www.thegreenwolf.com/animal-parts-laws. If you're making a purchase make sure it's legal to do so in the location you're at before money changes hands. Also see if there are any restrictions on bringing your purchase back with you if you live in another state.

International travelers will want to be even more careful, as you'll need to declare everything to customs officials before you get on the plane or cross the border. Again, check the laws thoroughly to be sure that your purchase is both legal where you're buying, and that you are allowed to bring it back home with you. Keep all receipts and relevant paperwork with the specimen, and if you have time before you head home print out copies of any relevant laws that support your purchase.

Packing It Carefully

If you're driving in a car the whole time, the biggest concern will be making sure that any hard objects like skulls and antlers are safely strapped down. The last thing you want is for something to come flying at you if you take a turn a little too quickly or, even worse, become a projectile should you find yourself in an accident. Even a pelt flopping over onto the driver's face can be dangerous. The trunk is the safest place for your specimens, whether driving or parked. There are all too many people out there who will smash your window while you're away from your car and take whatever is easiest to carry away, so if you have that expensive articulated animal leg you just bought sitting in the back seat in full view, some opportunist may decide to help themselves. The trunk may also be slightly cooler in hot weather; some specimens can handle a bit of heat, but others may not do so well, so use your own discretion.

Air, train and bus travel, where someone else may be handling your luggage, have additional considerations. You may wish to keep smaller specimens in your personal bags so that you can make sure they're alright. Be aware that in the US, TSA may search any of your bags, even the ones you carry through the security line with you, but at least this way you can both explain what they've just found and make sure they don't abscond with it. It's a good idea to check with the airline you're flying with first to see what their regulations are on transporting animal remains.[23]

---

23 In 2017, one hunter had to make a last-minute postal shipment after the airline he was flying with refused to allow him to bring an entire cougar carcass in his suitcase. No carry-on carrion allowed, it would seem. Sauce: https://nypost.com/2017/12/28/hunter-tries-to-sneak-cougar-carcass-past-tsa/

If you have to pack something in your checked luggage that's going into the freight area of a bus or train, make sure it's well protected from damage as your bags are tossed around like volleyballs. I pack as though the handlers are going to have very bad days and decide to use my luggage as a punching bag to take their frustrations out on. Delicate items get wrapped in many layers of clothing, and smaller ones may even end up boxed up before getting zipped into the suitcase. Softer items like pelts make great padding.

If you just aren't comfortable with either of these options, you may wish to just mail your dead critters home. Some shops will offer this as an additional service for a fee. Or you can go to a post office or other shipping center, buy packing supplies, and get your specimens ready for their own trip home where they will hopefully arrive in excellent condition. (More details on how to pack things in the next section.)

For moves, be mindful of how far you have to travel. You don't need to be quite as careful for a cross-town trip in a U-Haul versus heading across the country, but you still want everything to make it there safely. I find that plastic storage bins with some packing materials work pretty well either way for bones and other breakables; you may wish to tape the lids on, especially for longer moves as things may shift around. Make sure anything delicate gets packed on top of heavy things so it doesn't end up getting damaged. Hides are a little more forgiving, and I've actually used them as padding for bones, or stuffed them in plastic garbage bags and let them be squished into smaller spaces in the truck. If you are driving your own vehicle in addition to someone driving a truck, you may want to have anything particularly valuable in the trunk or back seat.

If someone else is moving your stuff for you, make sure you have a complete inventory of your collection before the move. All too often items may be misplaced by accident, or stole in worse situations. Make sure that your moving company has liability written into the contract. If something does end up missing, don't be afraid to contact the police to file a report, especially if the moving company tries to deny any responsibility.

Don't Bother the Other Travelers

Look, I get that there is a certain subsection of Vulture Culture that tends to be a little more edgy and attention-seeking, like running around in wolf headdresses in downtown areas of major cities, or wearing rat skull necklaces to mainstream weddings. While I know this can be spun as an educational opportunity or even advertisement for artists out in the open, when you're traveling in a conveyance other than your own vehicle it's best to not be quite so exhibitionistic. It's not just about avoiding squicking out the people who think dead things are gross, either, which by the way is just rude when you're sharing a confined space like a plane or train for a few hours.

There are also physical and logistical considerations, too. Use your common sense and courtesy. If you're trying to get a taxidermy caribou mount home and you don't have a car, call a taxi van or even rent a U-Haul van rather than trying to take that giant-antlered beast on the bus. Avoid anything having to do with fresh roadkill if you're too far from home to walk and don't have a car or a friend with a car to help you get it home, especially in the heat of summer. And any time you're traveling with a wet specimen, drain the alcohol out of the jar, wrap the specimen in cloth soaked in alcohol and put it back in the jar, then pack it carefully. All of this should help you avoid unfortunate public occurrences.

Downsizing Your Collection

As much fun as it can be to build up a big collection, the reality is that a lot of Vultures end up having to sell or otherwise get rid of parts of it from time to time. Sometimes it's a need for space; other times they need the money. Should you find yourself in this position, here are some options.

Sales

You've likely invested at least some money in your collection, so it's to be expected you'd like to get some of that investment back. You have several different options for online sales of hides, bones and the like; here are some of the more popular ones:

- eBay (http://www.ebay.com)
- Etsy (http://www.etsy.com)
- Taxidermy.net (http://www.taxidermy.net/forum/index.php/)
- Craig's List (http://craigslist.org)

Make sure that you review the animal parts laws section of this book, as well as my animal parts laws database at http://www.thegreenwolf.com/animal-parts-laws/ before you try to sell anything in your collection. You want to be sure that it's legal to sell in your country/state, and also that anyone trying to buy it can do so legally where they are. Also be aware that Etsy and eBay all have various restrictions on selling certain animal parts on their sites, and trying to sell prohibited species may end up with your listing being taken down or your account being suspended. eBay's list is in their policies, and Etsy's is in their legal section.

A good listing starts with good pictures. Even if you don't have a great camera, try to take the photos in good light; I like natural light outdoors, especially on an overcast day where the clouds act as a nice filter. If you're photographing indoors and don't have special equipment, at least wait til daytime and try to take the shot near the window. Poor lighting will just make the photo quality worse, and if you're trying to take pictures with a cell phone you want to try to make up for that as much as possible. Take pictures from a few different angles, and get close-ups of any flaws or other notable details.

Be thorough in your description. Talk about what it is, where you got it originally, any neat details or damage, and anything else you might have liked to know about when you were buying it. Include measurements like length and width.

If a particular item can't be shipped to certain places, note that in the description as well.

Pricing may be one of the toughest things about selling specimens. You can't go wrong with selling it at the same price you got it for. You can also look at other people's listings for specimens of a similar type and condition to get an idea of the range of prices they're going for. Be aware people sometimes seriously inflate their prices, and others will charge much less than what the item is worth, so try to go for an average.

If the item in question is a piece of your artwork made with fur or bone or somesuch, that adds another layer into price considerations. Think about the cost for your materials, and how much time you put into the piece, and compare your prices to those of other hide and bone artists making comparable artwork. If you're a more experienced artist you can get away with charging more than a brand-new artist, too. Some artists try to factor in at least minimum wage; this may or may not work, depending on the current price range for similar art. It is pretty complicated, and even after two decades of making hide and bone art I still sometimes have trouble figuring out the right price for something I've created. Be patient, allow yourself to adjust your prices as you go, and eventually you'll get a good feel for the balance between what you put into your art and what people are willing to give you for it.

Shipping is one more thing to keep in mind. If you haven't shipped very many packages, it's a good idea to use a postage calculator to estimate what shipping will cost; I like using the US Post Office myself. You can always charge a little more and then refund the extra back to the buyer. For padding, I've used everything from packing peanuts to plastic bags to newspaper, and I include pretty much every clean piece of plastic I get my hands on including toilet paper wrappers and plastic shrink-wrap from DVDs. For bones and other fragile items, you want to pack it in tightly enough that if you shake the box it doesn't move, but also be sure there's adequate space between the item and the sides of the box so if it gets dropped or kicked the padding absorbs the shock.

There are certain extra precautions to be taken with wet specimens, since they are often cured with fluids considered hazardous by the U.S. Postal Service and other shipping companies. You'll need to drain off the alcohol or other fluid, other than soaking some rags in it. Then wrap the specimens in those rags, seal the jar tight, and pack the jar very carefully. There are more details about shipping these specimens at https://www.nps.gov/museum/publications/conserveogram/11-13.pdf.

What if you don't want to sell online? Well, your options are a lot fewer. You can try running an ad in a local paper, though Craig's List is likely to get you more results. You can try calling a local taxidermist to see if they're interested, or if they do any consignment. Antique shops sometimes also consign, though often you have to lease a booth on a monthly basis. Some urban areas have oddity shops that specialize in vintage taxidermy and other specimens, and many are looking for new stock, either to purchase or consign. You could always have a yard sale, though don't expect a lot of results, and anticipate a lot of lowball offers.

Donations

If you don't necessarily need the money right away, you may be able to donate items to a local nonprofit like a wildlife center or other educational facility and get a tax write-off. Universities and museums may also be interested in very good-quality specimens, though many already have very full storage areas. Moreover, many such facilities also have very strict parameters for acquiring specimens, so if you inherited a leopard skin rug from a deceased relative, but it has no paperwork showing its provenance, they likely can't take it as there's no proof it was legally hunted other than your word. Still, it's worth calling around to local organizations and seeing if they're interested or have recommendations.

If you have items you know are legal, you might check with your local theater group to see if they can use them. Portland, OR has a secondhand art supply store called SCRAP which occasionally gets taxidermy pieces in, so if you have a resale shop for art supplies, furnishings or other odd stuff you might check with them.

Gifts

Most people love getting things for free! If you need to make some space but aren't particular about money, consider giving away some of your specimens to people you know will appreciate them. I tend to be very protective of my collection, so I won't give things away to just anyone; I want to be sure they'll be respectful and take good care of them. Handing off a wet specimen to someone who's grossed out by such things is a patently bad idea, both because they may just throw it in the trash, and because they may wonder whether you were just trying to be mean to them. Give that jarred critter to the right person, though, and you may be providing the best gift ever, whether it's their birthday or not! (When in doubt, ask the potential receiver of your gift whether it's something they'd like or not.)

Return to the Earth

If worse comes to worst, you can always put something outside to naturally decay and become a part of nature's cycles again. This really only works well with bones, feathers, pinned bugs and brain or egg tanned hides; you do not want to do it with anything treated with chemicals as they can leach into the soil and water. Burial may seem like the best option, but leaving them to decay on the ground's surface is actually better. Rodents and other animals like to gnaw on bones for the calcium, and other creatures may want to use feathers or hair for nesting material. And for all you know you could be starting someone else's career as a Vulture when they come across what you left behind and are inspired to find out more!

Chapter 6: Vultures and Our Neighbors

This chapter is dedicated to dealing with people and their varying relationships—positive and negative—to Vulture Culture. While I'd like to say everyone has a live-and-let-live (irony fully intended) attitude toward us, unfortunately some people think anything involving dead animals is horribly wrong

Bring Out Your Dead! Vulture Culture and the Media

The past few years have seen several articles in both print and online media related to Vulture Culture. These range from a National Geographic exploration of what it takes for a museum to maintain its taxidermy collection in good order[24], to the Smithsonian asking why more women are attracted to taxidermy than ever[25], to Elle's surprise at the rise in the number of vegetarians going to taxidermy school these days[26]. There are more books on the subject, too, particularly taxidermy; Appendix I has a recommended reading list.

I doubt we'll ever be more than a curious niche topic, but I for one am glad we've been getting a little more positive press. Since so many of us have spent time being ridiculed or harassed for collecting hides and bones, it's nice to have an article or book to point to and say "Look? It's not just me, and I'm not just some freak!" (I hope you might be able to do that with this book, by the way.)

Should you ever have a reporter contact you about interviewing you for a Vulture Culture article, I do recommend being a bit cautious. While the majority of articles have been good, there's always the chance you'll run into some bush league muckraker who just wants to get pageviews or angry letters to the editor with a harsh piece on these crazy people who run around with dead animals everywhere. Ask them what their intention with the article is, and what some of their possible questions may be. Check out some of their previous articles to get a feel for what they typically write. They may also want to take pictures of you, which means your face will be attached to this write-up.

If you decide to go for it, think ahead about how you might answer the questions. You'll likely get something similar to these:

- So why do you collect dead things?
- How did you get interested, and how old were you?
- What do you do with them?

---

24 http://ngm.nationalgeographic.com/2015/08/taxidermy/christy-text
25 http://www.smithsonianmag.com/arts-culture/why-taxidermy-being-revived-21st-century-180955644/?no-ist
26 http://www.elleuk.com/life-and-culture/culture/articles/a32430/how-taxidermy-became-cool-again/

- How do your family/friends/other people react?
- Do you kill the animals yourself/how do the animals die?
- What are your thoughts on animal rights activists who think what you're doing is wrong?

A good journalist will go into more personalized questions, like asking you your favorite piece in your collection, or if you have any recommendations for anyone else interested in Vulture Culture. No matter what the questions are, do your best to answer factually and politely. If it looks like the journalist is steering things in an uncomfortable direction, like asking leading questions about animal abuse and the like, end the interview politely but firmly.

Regardless of the tone of the final article, be prepared to be contacted by people who read it, whether aspiring Vultures or angry critics. Later in this chapter you'll get some advice on how to deal with the latter group; for the former, you may end up making some new friends! And if you've felt alone in your Vulture-hood, this isn't such a bad turnout at all.

On Finding Other Vultures

No person is an island, and that goes for us Vultures, too! One of the best things about Vulture Culture is that it brings together a variety of people who all share an appreciation for dead stuff. If you don't already have friends who share your Vulture-ish ways, the best way to find some is to start online. The two sites I use the most for this are Facebook and Tumblr.

Facebook has a number of groups dedicated to Vulture Culture; some are sales groups, others are strictly for discussion.

Please do check your privacy settings before joining any groups if you're at all concerned about family, friends or employers knowing about your interests. I mean, if you don't mind fielding a bunch of questions about why you're on a group called "Dead Things For Sale or Trade" by all means leave your privacy settings wide open if you like. Otherwise, here are a few tips for tightening security:

- Be mindful of who you add as a friend. You likely won't add your employer's personal profile, but sometimes there's no ignoring nosy Aunt Martha and Uncle Joe when they send you friend requests.
- If you do have to add people who may be weirded out by Vulture Culture, make your Vulture-related posts viewable only by a select group of your friends. There are two ways to do this. When you go to make a post, you can choose the "Friends except..." option in the privacy dropdown menu and then choose whoever you want to filter out. This is the best option if there are only a few people you want to keep out of your Vulture business. If your Vulture circle is relatively small and you need to maintain privacy from a bunch of people, choose the "Specific Friends" option. If you need to filter the same people in or out on multiple posts, make lists of Vulture-

friendly and Vulture-unfriendly people. To create a custom list of people, go to "Explore" on the left side of your news feed, and then click "Create List." Then add whoever you want to be on that list, and when you're done click "Create."

- If you don't want people seeing what public groups you're a part of, go to your profile. Click on the "More" dropdown menu near the top of the page and then click on Manage Sections. Then make sure the box next to Groups is NOT checked.

- Speaking of groups, if you make any posts or comments on an open group (one where anyone can see the posts regardless of whether they're a member or not) these can show up on your friends' news feeds as "so-and-so posted in such-and-such group" or "so-and-so replied to a comment on a post in such-and-such group." Many Vulture-related groups are closed or secret so posts and comments that you make there shouldn't show up in your friends' feeds, though I have heard of occasional glitches where closed group activity *did* show up. Be aware.

Keep in mind that the privacy info I have here is current as of the publication of this book, and Facebook likes to periodically switch around the layout of the site so that you may have to look for certain features in a new location.

Tumblr is a much simpler matter. Not only can you use whatever name or handle you like (which is good as Tumblr is a lot more public), but there aren't groups or pages. On the other hand, there are fewer central hubs for finding like-minded people. The best thing Tumblr has to offer in that regard are tags, and the two best I've found are #vulture culture and #taxidermy. If you see a post with one of those, or another that looks promising, click on the tag and it will take you to a feed of tagged posts. Or go to https://www.tumblr.com/tagged/vulture-culture or https://www.tumblr.com/tagged/taxidermy.

Once you're there, you're welcome to start following the Tumblr blogs of people posting in those tags. They don't have to follow you back in order for their posts to show up on your dashboard. I recommend reblogging posts that you like, and leaving nice comments as well. Some Vulture-related blogs even allow followers to ask questions about Vulture Culture.

You can also use tags on Instagram and Twitter to find like-minded people. I don't personally like these platforms as well as Facebook and Tumblr, but your mileage may vary.

Vultures in the Flesh!

Online socialization is great, but some people really prefer in-person connections. If you want to meet Vultures in person, here are some ideas.

- If you have a local oddity shop, see if they host any classes or events relating to taxidermy or other Vulture-friendly topics. Not only will you have the opportunity to learn something, but you'll also get to meet other people with similar interests. If you feel confident enough, you can even see if you can present a talk or workshop of your own there!

- If there are other Vultures online that you've known a while and they live in your area, you may consider inviting them to meet in person. Of course, you should follow basic "meeting people from the internet" safety practices, like going with an adult if you're a minor, and meeting in a public place first. Who knows? This may be the start of a wonderful in-person friendship!

- If you want to be really proactive, start a Vulture Culture club in your area! Find a good public place like a coffee shop to meet, pick a day and time (weekday evenings are especially good) and then promote it online in various Vulture Culture groups and other places. You can also put up posters on local bulletin boards. If you have an oddity shop or similar business in your area, they may even be willing to host the club meeting if they have space, or at least put up a poster. Libraries also often have meeting space available, though you may have to reserve it ahead of time. The first meeting is likely just going to be everyone getting to know each other, and that's okay! You can decide whether you want it to keep being a casual meeting every month or so, or whether you want to be more structured with speakers, classes and the like. I recommend starting out small and simple, and seeing where things grow. In between meetings you can keep in touch with an email listserve, Facebook group or similar online presence; I recommend not making it public so as to protect the identities of members who may not have as much support for their Vulture-hood as others.

- Depending on where you are you may be able to arrange a group outing to look for bones and shed antlers. Sharp eyes win the prize! However, be aware of any federal or state laws; for example, you cannot take any natural items from U.S. National Parks. Some states have restrictions on when you can or cannot go looking for shed antlers.[27]

You may also find that you run into fellow Vultures in other settings. I've found them at sci-fi/fantasy/anime/etc. conventions, goth clubs, art shows, nature organizations, and even at parties thrown by mutual friends. If you're open about being a Vulture, you'll find that after a while people seek you out or are directed toward you by well-meaning friends wanting to help create a connection.

Unfortunately, not everyone who finds out about your interest in animal remains will be amenable to it. Let's talk about that next.

---

27 http://www.bonecollector.com/western-public-land-shed-hunting-laws/

Dealing Diplomatically with Disagreement

You can generally divide people who disagree with you into two categories: other Vultures who disagree on certain nuances of Vulture Culture (such as hunted vs. natural death hides), and non-Vultures who think this whole dead animals thing is wrong. The first group can sometimes be as difficult to deal with as the second, as people do like to hold their ground when it comes to opinions. It may be that they genuinely feel that you're causing harm through how you obtain your collection, or they might just be trying to stir up drama in an attempt to gain popularity. At least you can still try to find some common ground with them in your shared membership in Vulture Culture.

Now, not everyone outside of Vulture Culture is going to be offended by it. I have been selling my artwork in person and online for many years, and the vast majority of people who have questions or concerns about it are nice and polite, and willing to listen to my answers and explanations. In fact, I've had some really good conversations with people who have agreed to disagree with me.

People who really hate Vulture Culture (we'll call them Anti-Vultures) are a different story. For various reasons they think that anyone who likes hides and bones is sick and wrong. They might be convinced that all of us are secretly serial killers in the making, mistakenly confusing us for people who enjoy torturing live animals to death. However, they may also rightfully point out that some of the hides and bones being traded around do some from animals that were raised on fur farms or factory farms, and therefore feel that we are supporting inhumane industries. On rare occasion you may run into someone who has a significant phobia of taxidermy and other dead things, whether due to some sort of associated trauma or an unusual psychological quirk.[28]

If Anti-Vultures stuck to civil discourse that would be one thing. However, most Vultures have at least one story, and usually more than one, of people screaming at them online (or less often, in person) about how horrible they are. Some of the things I've personally been told are alternately obnoxious, disturbing, and occasionally nonsensical:

- That's sick!
- You're going to make coyotes go extinct!
- You should be banned from [insert website here]
- You're sick and should be locked up!
- MURDERER!
- Maybe someone should kill and skin you!

---

28 While I was writing this book, an opinion article came out in one of our local weekly newspapers by a woman who has a serious taxidermy phobia brought on by a couple of childhood traumas. Unfortunately, her attempt to try to understand why taxidermy is such a common "quirky" decoration here in Portland only resulted in her spending several hundred words projecting her fears onto an entire city. You can read it archived at https://www.portlandmercury.com/feature/2018/01/17/19610002/overkill

- Don't you know they kill the animals while they're still ALIVE?
- [After getting a reasonable answer to their question or concern about fur/bones/etc.] Well, I still think you're wrong!

I've also had people criticize my art while wearing leather or eating meat (like hot dogs at street fairs I've vended at), or while wearing non-biodegradable fake fur or pleather made from petroleum. I used to try to reason with them or get into debates, but as I've gotten older and more tired I discovered just how utterly pointless it is. Once someone's got their hackles up, they're generally not going to listen to any other perspectives than their own. In this case, my own tactic is to not engage if I can help it.

If I do need to deal with someone like that, such as at an in-person event, I try to be as polite and friendly as I can, and de-escalate the situation if possible. I'll also say "Hey, I think this may not be the booth for you; can you please leave?" Generally they head out at that point, though I have had a couple of scary situations where it took more firmness—and event security—to get them to stop harassing me.

That's one of the big reasons I prefer de-escalation and disengaging as quickly as possible, both online and in person. It's not just a waste of time and energy—it can occasionally get threatening. I have a file on my hard drives of screen shots of threatening emails and comments I've received, just in case I ever need to go to the police with them. And in person I never know whether the Loud Yelling Person in my booth is suddenly going to turn into the Physically Violent Person. (Just because I have martial arts training does not mean that I want to have to use it.)

Moreover, I think it's best to take the high road. There's a pretty vocal blog/website supporting the fur industry, Truth About Fur (https://www.truthaboutfur.com/). I follow their social media because sometimes they post some really neat links on people in the fur trade like traditional trappers, or a series of blog posts following the biodegradation of a piece of real fur versus a piece of fake fur (spoiler: the real fur disintegrated within months, while the fake fur is still intact.) However, sometimes they post articles about animal rights activists and vegans, and their tone is pretty antagonistic and insulting. I would like them a lot better if they did only the former sort of posts, and not the latter.

We want to set a good example for others and show the rest of the world that we're just everyday people with an interest in one particular slice of nature. Part of how we can do that is by being polite, informative, and diplomatic in dealing with everyone, whether it's a curious person with a question, or a raging Anti-Vulture on the attack. Getting pissed off and yelling back or writing a vitriolic reply may feel better in the moment, but it's just going to train you to respond to attacks with anger, and probably make them more antagonistic, too. It also makes you look just as bad as they are. Better to let them be the sole source of upset.

If you do have a run-in with Anti-Vultures, once you've managed to see them off, spend some time in some soothing self-care to calm yourself down, and then do your best to move forward and leave the unpleasantness in the past. There's so much more to your life than what a stranger on the internet thinks of you, and they don't deserve that much power over you.

This can be easier said than done, of course. There have been cases of Vultures being so badly harassed and threatened, especially online, that it's caused serious mental distress. I hope this never, ever happens to you, but if it does, please get help. Ask a trusted friend for support and a shoulder to cry on; you may even ask them to go onto the account where you're being harassed and delete nasty comments and messages after saving screen shots. If it's really bothering you, consider seeing a therapist to help you deal with the trauma (yes, even online harassment can be legitimately traumatizing!)

In the rare event that the harassment turns to serious threats—for example, someone online telling you that they're going to kill you—screen shot the threat and take it to the police. The same goes for someone threatening or stalking you in person. I can't guarantee you that they'll take it seriously, because unfortunately not all police departments have caught up to the severity of online or in-person threats. But you can at least start a paper trail of reports. If you do not have a lawyer and can't afford one, you may also wish to contact a legal aid organization as they specialize in offering legal advice and support to people with low or no income.

Being Honest With Ourselves

As much as we may dislike being screamed at by Anti-Vultures, sometimes they do have something of a point. Remember earlier in the book where I talked about how some Vultures don't like buying or owning remains of animals that were raised on fur farms? Most of the time when I have someone vociferously trying to dissuade me from my Vulture-ish actions I do my disengage routine and let them go on their un-merry way, because they're usually screaming the same thing as every other Anti-Vulture out there.

But at times I do have cause to stop and think. It's a good idea in general to re-evaluate one's ethical stances now and then, and so I contemplate what my feelings are about the fact that I am collecting and making art with the remains of animals who may not have had good deaths. I didn't kill them myself, but that doesn't mean I've had no impact.

The fact is that yes, fur farm animals live in small cages for the entirety of their lives. That's not good for animals who traditionally had thousands of acres to roam on. Even if fur farm animals are of strains bred for captivity, captive animals would still be happier having more space and enrichment. And I have to remember that every time I work with a hide from a farmed animal.

I also have to be honest about how just because a species is legal to buy and sell doesn't mean that their hunting is done sustainably. Up until 2016, it was legal to buy and sell African lion remains in the US, until they were added to the Endangered Species list. In the sixty years preceding that listing, lion populations in the wild dropped from over 400,000 to about 20,000. And there are many more species, especially outside of Western countries, where we don't have enough information to know whether hunting and/or trapping significantly affects their numbers.

While individual Vultures may limit themselves to roadkill, it would be wrong to say that Vulture Culture as a whole is free of ethical conundrums of this sort. And that gives Anti-Vultures fuel because to them, there's only one real answer: no animal parts at all. (Though some of them may disagree on whether meat is okay, but that's a whole different story.)

How we answer these ethical issues varies from individual to individual. My own solution is to be mindful of where my various hides, bones and the like come from, and to try to acquire secondhand or remnant scraps as much as possible. I buy a fair amount from people's private collections, and while I don't have as much time to wander around fields and forests looking for bones, I do still manage to come across a treasure trove every once in a while. I also try to counteract my impact by donating some of the money from my art sales to nonprofit organizations that benefit wildlife and their habitats. I know it's not a perfect solution, but it's the one that's worked out the best for me so far.

Vultures need to understand that the remains we work with didn't just materialize out of nowhere. They came out of ecosystems, even if some of those ecosystems are heavily human-altered. We have the opportunity to make something bigger out of Vulture Culture, and that's what I want to dedicate the last chapter to.

Chapter 7: Rewilding the Beast: Vulture Culture as a Return to Nature

While Vulture Culture is by and large about collecting dead stuff, it is also intimately linked with the natural world. Unfortunately in Western cultures in particular we've gotten really good at thinking we're disconnected from the rest of nature. So it is entirely possible for people to acquire hides, bones, feathers and the like without ever really thinking about the animals and ecosystems these came from.

A lot of this is because modern society is so incredibly convenient, and people are so detached from the sources of our food, clothing, shelters, entertainment, etc. that we simply don't need to think about such things. Think about it: how often do you think about the chicken that laid the egg you eat for breakfast, or the seamstress who sewed the shirt you're wearing, or the miner who dug out the precious metals in your cell phone? Chances are you plunked down money for the egg, shirt and phone without giving a single thought to their supply chains.

The trouble with this, of course, is that it's easier for the companies behind these supply chains to hide abuses and other damages that result. Most people don't especially care that their egg came from a bird stuck in a cage for its entire life (or that most "free range" eggs are from chickens crowded on the floor of a dark, smelly barn.) Nor do they really concern themselves with how little of their money goes to the sweatshop seamstress or miner working in dangerous circumstances. And if so few people care, who's there to hold the companies accountable?

The same goes for the environment. Part of why there's such widespread deforestation, pollution, and extinction is because we developed our technologies without really considering the negative impacts on nature. And now that we recognize them, it's incredibly hard to get people and institutions to change their ways, especially those entities who have a financial stake in the status quo. Look at how many people still deny human-caused climate change because they want to keep making money on fossil fuels or clearcutting forests.

We can find the root of this selfishness in our perceived detachment from the rest of nature. When you think that you aren't a part of a system, you're less likely to care about what happens to it. We've made up all sorts of stories to try to distance ourselves from nature, whether religious (Genesis 1:26 and its "Man shall have dominion over the Earth" message), philosophical (mind-body dualism, which mirrors the idea that "smart" humans are separate from the rest of the body of nature) or technological (our increased tendency to stay indoors and distract ourselves with electronic toys because nature is supposedly dirty and unsafe.)

Not only has this all had a negative effect on nature as we destroy it without a second thought, but it's had a really bad impact on us. We're spending less time outdoors than we used to because we lack motivation to do so. More children and adults than ever are showing the effects of Nature-Deficit Disorder, such as an inability to pay attention or sit still, increased stress and other mental health issues, poorer physical health, and less respect for nature (human nature included.) In short, this whole lie that we are separate from, and somehow better than, the rest of nature, is a huge disaster for everyone and everything on the entire planet.

I think Vulture Culture has the potential to counter some of this detachment. Yes, it's nice to appreciate the aesthetics of hides and bones, and the artistry that goes into a well-prepared specimen. It's even okay to enjoy the fun of collecting the skulls of all the legal species of a given genus of animals, or one of every color of fox hide. But if that's all you care about, you're falling into the same detached materialism of the rest of society, and I know we can do better than that.

How? Allow me to give you some specific possibilities to play with.

Vulture Culture as a Fascination with Natural History

When you collect or otherwise work with hides and bones, you are handling a deeply intimate part of nature, the remains of a once-living being. It's a very physical connection, especially when it's from an animal that you likely wouldn't be able to even get close to when it's alive. And this tactile experience is an invitation to learn more. Many people are shocked when they come into my booth and start touching the hides of wolves and foxes and bobcats because they had no idea they'd be so soft. Others are surprised at how large or small some of these animals actually are, especially since they may have only ever seen them at a distance in person, if at all.

We Vultures are not immune to these fascinations. In fact, many of us take the time to learn about the species whose remains we own—where they live, what their life cycle is like, why the bones are structured the way they are. We may learn the different parts of a skull, or how to tell what an animal eats just my looking at its teeth. Some of us enter into relationships with smaller beings in nature when we use dermestid beetles or bacteria to clean bones, and have to be able to understand and attend to their needs.

But we can go further than this. One of the threads contributing to Vulture Culture that I mentioned earlier in the book is the cabinet of curiosities, a collection of natural history specimens used for both personal education and shared scientific knowledge in the times before public museums. While there were certainly wealthy people who just had nicely stocked cabinets as status symbols, the practice was begun by those who were genuinely interested in studying nature in depth.

We have so many more resources than our predecessors did. Not only do we have more physical access to a wider range of ecosystems through travel and tools like microscopes, we also have more knowledge about their details through books, websites, classes and talks. We can visit national and state parks, refuges and wilderness areas and learn about them through rangers and other interpreters, educational displays and on-site bookstores. We would be foolish not to make the most of those as we're able.

I prefer a shift in attitude. Vulture Culture is still a heavily aesthetic movement; people like pretty things and they like the opportunity to acquire them, and that's okay. But I also want to see it more widely reframed as a natural history movement, one in which people are inspired to learn more about the nature around them through the specimens they collect. I want to see Vultures who can explain the basic life story of any species in their collection, and who can identify the sort of ecosystem this animal likely came from. I want to encourage Vultures to look at their collections not just as rustic décor or stuffed animals for grownups, but as a microcosm of the world around them.

Vulture Culture as Acceptance of Death

Just as Vulture Culture has the potential to bring us up close and personal with nature, so it also can help us come to terms with one of nature's most universal and inevitable phenomena: death. American culture in particular has a very unhealthy relationship with death; we seem to almost deny that it exists, instead being obsessed with trying to preserve youthfulness as long as possible. Our elderly are shut away if they can't keep up. Our funerary rites are solemn affairs with little space for open grieving and community support. Our jobs give us maybe a day to attend a funeral of a lost loved one, and then we're expected to be back on the job, right as rain.

Anyone who explores death beyond these tight confines is seen as morbid and weird, possibly to be avoided. Certainly there are people who have pathological obsessions, whether disturbed individuals who kill animals (or, occasionally, humans) for fun, or those whose anxiety with death is so severe that they can barely think of anything other than their mortality. But there's not a lot of room in between. Movies, books and other media that deal with death head-on in an emotional manner are seen as deep, especially when compared to countless action movies and shows where scads of nameless bad guys are killed on screen within seconds.

Our distance from nature also contributes to our discomfort with death. I've lost track of how many omnivores I've met who have neither killed their own meat nor attended a slaughter. Many even seem to be completely unaware of how an animal is processed into meat and other products. Considering animal products are everywhere from on our plates to our clothing and makeup to our medicines and beyond, we would do well to familiarize ourselves more with just how much animal deaths touch us on a daily basis.

Vulture Culture brings us in direct contact with that death. While many of the animals whose hides and bones we collect are inedible by the time we get them, we are under no illusion as to whether something died in order for us to have them. You can't deny mortality when looking into the empty eye sockets of a skull held in your hand, or running your fingers through the long hairs on a tanned pelt.

Accepting death doesn't mean wearing black clothing, reading *The Complete Works of Edgar Allen Poe*, and listening to the Cure and Siouxsie Sioux. (Not that we don't have our fair share of goths in our ranks, of course, me included.) For me, it means coming to a sort of peace with the idea that just as there was a world without me for billions of years, there will soon again be a world without me for billions more years, and in between there's the tiniest eyeblink of time in which I exist. And it's the same thing for every single living being that ever has been, is, and shall be on this planet. I am not particularly invested in the idea of an afterlife. If it's there I'll find out when it's my time to shuffle off the mortal coil.

However, the fact that we each only get such a miniscule amount of time in the grand scheme of things also brings more value to each life. I remember that whenever I handle animal parts that I am holding the remains of a once-living being who, like me, breathed and moved and ate and slept. When that being died, its unique view on this world also ended permanently, never to return. It's why I don't take death lightly, and on the occasions I kill a being for food, particularly an animal, I am mindful of the fact that there will be one less perspective to explore the universe. That, to me, is reason enough to be respectful of the dead, even if they no longer need their remains.[29]

Suffice it to say, every person has to come to terms with death in their own way, and that goes for Vultures along with the rest. By respectfully handling the remains of our dead non-human community, we are reminding ourselves that we, too, are flesh and bone and blood and sinew. And we remember that our bodies are only temporarily borrowed from the earth through the food we eat, and someday we'll have to give it back so others can make use of the tissues and molecules. I've planned a green burial for myself when I die, because I want my body to go back into those cycles as quickly and directly as possible. It's just one way for me to reconnect myself to nature.

Vulture Culture as a Form of Rewilding

I also want to see Vulture Culture become an avenue for us to rewild ourselves. Rewilding has several definitions, but in this case I'm referring to the practice of undoing domestication, in humans and otherwise.[30] Rewilding may include returning agricultural or other human-dominated spaces to their previous wilderness states, but it also includes remembering who we were as a species before the majority of us became civilized and domesticated.

One of the major efforts in rewilding is the learning or re-learning of archaic technologies (often inaccurately called "primitive" skills despite their technological sophistication.) Some of these may be very familiar to Vultures who

---

29 And for me that's where Vulture Culture becomes intensely spiritual. I am a naturalist pagan, which means I don't have any beliefs in the supernatural, but I do hold to personal ideas about archetypal beings and nature spirits. Even if they only really exist in my head, that's enough of a framework to contribute to the awe and wonder I have at the world around me. As I mentioned in an earlier chapter, I already wrote a book on the spiritual aspects of Vulture Culture so I won't elaborate upon it again here.

30 Peter Michael Bauer, also known as Urban Scout, is a pioneer of the rewilding movement in the U.S. If you're interested in this concept, his website at http://www.petermichaelbauer.com/what-is-rewilding/ is a good starting point.

process their own remains through hide tanning and bone cleaning. Other relevant skills involve crafting items like clothing and tools from these hides and bones.

Rewilding in Vulture Culture isn't just about who can create the coolest hide scraper from an elk antler, though. It's also about heavily questioning the effects civilization has had on us and the rest of nature. Hard-core rewilders feel that civilization is in the process of collapsing and that only those who have archaic skills will be able to survive by reverting to hunter-forager lifestyles. Others are less pessimistic about civilization's long-term survival, but see rewilding as a practical critique of our over-dependence on our technologies and our schism with the outdoors. As has already been established, our disconnection from nature has had bad effects all around, a problem rewilders are seeking their own solutions to.

What I would ask potential rewilding Vultures is: How can our interest in dead things lead us to wider discourses about technology old and new, and induce us to learn skills we have taken for granted? For example, look at how much Vulture Culture overlaps with do-it-yourself culture. Chances are if you've ever taken the time to tan a hide by hand you have a greater respect for the art and craft of it than you might have had before, because you know how difficult it is. When we have a greater appreciation for the effort and resources that go into manufacturing everyday items, we can begin to see how doing these same processes on massive, global scales can have immense consequences for people and planet. Granted, tanning a few dozen hides in a commercial process is more efficient in some ways and certainly takes up less time overall. But it results in more pollutants hide for hide than, say, brain tanning individual pelts.

What happens to us and our society when we are able to become less dependent on mass-manufacturing and conventional agriculture when we are able to produce our own food and goods from the animals (and plants, and fungi) around us? We begin to question the overarching systems that are in place which keep us insulated from their inner workings and discourage us from thinking about where everything come from. We also see the urgency in preserving not just archaic skills, but the indigenous people who still rely on them in the 21st century, many of whom are threatened by the encroachment of corporate interests and civilizations that often treat them violently. And we look for more local, sustainable alternatives to a "business as usual" future.

I'm not saying that you have to go out and hunt deer next fall so that you can have a nifty new brain-tanned outfit and replace all your beef with venison, though you're certainly welcome to do so if you have the time, resources and ability. Just questioning the status quo through a rewilding lens can be valuable, especially if it causes you to be a more conscious member of your community, human and otherwise.

Vulture Culture as Conservation Movement

I hope that by this point in the book I've made it clear that not every bit of hide or bone out there was got through painless means, nor are we innocent of contributing to species endangerment and potential extinction. But we have the potential to be a movement that gives back to nature at least as much as we give.

Some of that includes that deep soul-searching regarding ethics. It's imperative that we all be as informed and honest as we can with ourselves about the actual impact of our collections, and try to live as close to our personal boundaries as possible. For myself, I recognize that some of the materials I work with ultimately came from hunted, trapped or farmed animals, and that I choose to work with them knowing that. My reason is that I want them to go to good use rather than being wasted, which is why I make use of even the tiniest scraps, one way or another. And I revisit that decision periodically to be sure I can still live with it.

I talked earlier in the book about how just because an animal is legal to possess, doesn't mean that its remains were acquired in a sustainable manner that will support the ongoing existence of the species. It is imperative that we look more deeply into this issue than just the laws on the books. Anyone who has paid attention to conservation laws knows that it can take years, if not decades, for a law to be passed that effectively protects some endangered species. In the United States I have watched for a long time as opponents of these laws have tried to have them reversed so as to allow for more logging, mining and other destructive practices in the homes of protected species. So the laws are a start, but not the ending.

The International Union for Conservation of Nature has spent the past fifty years monitoring animal and plant species and assessing how close they are to extinction. The criteria range from Least Concern to Extinct, with several levels in between. There are also many species that are Data Deficient, meaning there's not enough information on them to know how threatened they are; I tend to treat Data Deficient species as endangered, just to be on the safe side. Wikipedia includes IUCN info on the pages for many species, making it easy to see their current status. I would urge my fellow Vultures to start using the IUCN Red List as a second guideline (in addition to laws) as to whether to buy/sell/trade a given species' remains or not.

Not every country's hunting laws are as well-defined as those in the United States, and some may have little to no regulation, particularly for non-big-game species. There is a seemingly endless list of animal, particularly smaller ones like birds and monkeys, whose numbers are in steep decline due to over-hunting even though that hunting may be technically legal. Given that many specimens are of dubious provenance, particularly those not from countries with extensive laws, we should be asking more about where these specimens come from, how they were killed, and how badly local populations, if not the entire species, are affected.

It's also really important to educate ourselves on the issues that affect not just wildlife but their entire ecosystems. Climate change is the most globally relevant issue, but habitat destruction through other means is the biggest cause of species endangerment and extinction. Pollution, development, and introduction of invasive species are just a few of the ways in which we humans have managed to damage ecosystems.

And it's important for us to critically examine some of the dialogue around Vulture Culture with an ecological eye. For example, I frequently see people justifying the killing of coyotes, deer and other widespread animals as "population control." Very seldom does anyone mention that these animals are overpopulated in many areas due to the local extinction of large predators like wolves and mountain lions that normally keep them in check. (And when large predators do reestablish themselves in their historical range, with or without human help, there's often a cry to kill them off so that we have more deer and elk to hunt—yikes.) Furthermore, the destruction of wilderness to create suburbs actually creates more favorable deer habitat and food sources, leading to population explosions. And coyotes have even evolved to have larger litters of pups if the local coyote population drops, so all "population control" does is encourage more breeding.

I've also seen the idea that the answer to human-animal conflicts is to kill the animals. Truth About Fur's social media regularly posts about animals inadvertently causing destruction to human habitation (such as beavers causing flooding with dams), and their answer always seems to be "this is why we need trappers!" (http://www.truthaboutfur.com/blog/value-of-an-active-trapper/ is one example.) I've seen this attitude repeated on some of the Vulture Culture related groups and forums I'm a part of online, and as a naturalist I find it pretty ecologically incorrect, as well as personally distressing.

Why is this important to Vulture Culture? Because if we do not act responsibly toward the world around us, we're just another group of materialistic humans doing nothing but take, take, take. If we care about nature as much as we claim we do, we need to back that up with action. From a more selfish but practical perspective, if there's nowhere for wild animals to live and die, there's nowhere to go looking around for their bones, and no bones to be found anywhere.

Vulture Culture as Environmental Education

You know what's great about acquiring all this knowledge about natural history and rewilding and ecosystems? Sharing it! We have a serious deficiency in nature literacy in Western societies, which again contributes to our tendency to destroy nature even without intending to. By educating others in constructive, meaningful ways we can help improve nature literacy and help others reconnect with and protect the great outdoors.

Vulture Culture can be something of a touchy topic in environmental education. Many nature lovers have trouble with obviously dead animals, especially when the face is still attached. Even people who will happily buy meat from the grocery store may recoil at the sight of a pelt. By moving Vulture Culture toward a more aware and responsible direction, and by pointing out that a lot of Vultures are environmentalists and conservationists, we can start showing other people that we do care about nature as much as they do, and that we aren't just filling our homes with dead things as trophies.

Many people see synthetic alternatives, which are often made from petroleum and are not biodegradable, as more "eco-friendly". Let's make one thing clear: hides and bones are not without ecological problems, both in the deaths of animals and in how many hides are commercially tanned. But at least they will biodegrade in a few years, whereas plastics can persist for centuries, breaking down into smaller and smaller pieces that are eaten by animals that then die awful deaths. So we have the opportunity to promote animal-based natural materials as environmentally sound, especially if they are acquired and processed through more sustainable means, particularly by individuals.

On the other end of the spectrum, there are hunters and other outdoor enthusiasts who are put off by the term "environmentalist" and expect any tree-hugger to attack them as evil animal-slaughterers roaming the land in search of Bambi. Since many of them own taxidermy, this can serve as a piece of common ground to show that we're not interested in getting into a fight. In fact, many outdoorspeople are conservationists who want to protect nature as much as we do, even if our reasons ay sometimes vary, so our collections can be good talking points to start with.

Regardless of the audience, I would like to see Vultures doing more outreach and education. We already have a pretty impressive body of knowledge, from animal anatomy to wildlife laws to preservation methods for hides, bones and more. If we offer that to other people, we are providing yet another avenue for all of us to reconnect with nature.

A Few Ideas For Outreach

If you've read any of my previous works, you've probably noticed I'm a huge fan of helping environmental and conservation groups through financial donations and volunteer time. Since the 1990s I have donated a portion of the money I make from my hide and bone artwork, as well as my books on nature and nature spirituality, to these sorts of organizations. You don't have to be rich to be able to make a difference. Even if you choose to give a monthly donation of $5 to one single organization, it's a good effort.[31] Check with your workplace to see if they'll match your donation, or run a small fundraising effort and invite friends and family to participate.

I have spent many years actively volunteering to restore local ecosystems and protect them from further damage, as well as educating the public on their conservation. I also became an Oregon Master Naturalist as a way to gain more skills and opportunities in this regard, and to learn more about the places I want to protect. You might not be able to make it to every tree planting or litter pickup, but even one Saturday morning spent helping to remove invasive plant species so native ones can be put in their place is a great gift to nature.

One of the best volunteer opportunities is the aforementioned environmental education. We already have great visual aids in our collections. Bones, pelts, feathers and other specimens aren't just great conversation starters— they're also ready-made diagrams for everything from explaining the difference between herbivores, carnivores and omnivores, to showing how an animal's fur protects it from the elements. Vultures have taken their collections to schools, scout meetings, even libraries to educate people about both taxidermy-related topics and

---

31 Don't forget that Appendix II has a list of nonprofit organizations I recommend!

natural history in general. If you want to do something similar, contact somebody in charge of the group/institution and let them know what you're offering to do for them. Then if they accept, plan out what exactly you want to talk about. It's great to let people explore the specimens you bring, but also have a presentation prepared so that it's more formal than a simple show-and-tell. You can talk about Vulture Culture, about the structure of skulls or how hides are tanned, or even about the natural history of the animals whose remains you've brought. Be careful when letting other people touch your collection, especially younger kids; you might institute a "two finger" rule, where they can only gently touch something with two fingers. Another possibility is if you happen to have a pet, like a snake, try bringing in both the live snake and a few snake skulls, skeletons and/or skins, and do a presentation on serpents in general.

If you aren't comfortable being out in public like that but still want to help, consider writing articles for websites and other publications on Vulture Culture and related topics. You can write how-tos for sites like Instructables, or more journalistic "here's who the Vultures are!" pieces for news outlets. If you aren't an experienced writer you might want to start with smaller or more local publications, or even just starting your own blog. Blogging is especially great as you have more creative control, you can post as often as you want, and you can have as many photos, videos and other supplementary materials as you want.

Sometimes the best education happens organically. It may be a matter of a friend of family member asking you about your collection and giving you the opportunity to tell them more about it. Or you could find yourself in a conversation about taxidermy, hunting or other Vulture-related topics, and be able to share your expertise. Some people like correcting misinformed animal rights activists and other people, especially online, but again this can often result in pointless arguments. So choose your battlefields carefully! Don't feel like you're too new or inexperienced to be able to have a voice. It's okay to say "Well, I don't know everything on the subject, but here's what I do know." (And hey, you can always direct curious people to this book for a good overview!)

While volunteering is an excellent pursuit, its also okay if you manage to make at least part of your living with paid environmental education. Most of us won't be able to find full-time jobs doing Vulture-related things, but you might be able to swing a side gig here and there. As an example, for a few years I ran a drawing session in Portland, OR called Still Death; artists got to spend the session drawing, painting and otherwise using a selection of my skull collection as art reference, and I also gave a talk on some element of comparative anatomy partway through the session, using the skulls to illustrate my points.

All of this outreach serves multiple purposes. It improves the general public's view of Vulture Culture, which is especially important since there are still some misinformed folks out there. It also helps people who may be interested in what we do to find a good starting point for getting involved. If you're branching out beyond strict Vulture Culture topics into more general natural history, you're benefitting nature on a variety of levels. You could make some new friends and connections. If you start doing this sort of education on a regular basis, even if it's just as a volunteer, it looks good on a resume, too.

These are just a few of the ways we could enrich and develop Vulture Culture into something deeper and wider-spanning than just "we like dead things!", focusing especially on natural history and reconnection with the outdoors. There are other directions in which we might expand our community, from exploring our artistry in more detail, to delving deeper into why Vulture Culture is so female-friendly and whether it might even be downright *feminist*.

What frontiers do you feel Vulture Culture could explore in the years to come?

Conclusion

I have been privileged to see this community grow from its beginnings into a beautiful and diverse group of people united by curiosity, creativity and a love of nature, and I hope that we can be more enduring than a mere fad. I want to see us be a community that can support each other and the planet we live on with a wealth of other beings. I want us to be able to educate others and share our curiosity, not just about dead things but about the living things, too.

For over a decade I have found a kinship with other people who are, like me, sometimes a little odd in our interests, but passionate about them too. I've learned so much about everything from bone cleaning to animal parts laws to conservation from people who are eager to share what they know. Most importantly, I've learned that I'm in good company in my fascination with hides and bones. As a child I sometimes felt strange and isolated in my explorations of the natural world; as an adult, I know I was never really alone at all. And that's what I hope my readers know, too: that this is your invitation to our community, and you're welcome to join us.

Vulture Culture was one of the threads that I followed in my adulthood to rediscover the wonder and awe that I developed toward nature as a child. I have a wealth of knowledge and understanding I couldn't even have imagined then, and I also feel a deeper sense of responsibility for the planet, even as habitat loss and other pressures push species to extinction, and climate change threatens to bring an end to *Homo sapiens sapiens* as well. I'm no fan of denying serious issues, but I enjoy a bit of escapism now and then as look at my collection of animal skulls, comparing the way the same structure evolved in different species, and how each of these uniquely solved the problem of surviving from one generation to the next.

And yes, on the shelf there's a little cottontail skull, perhaps a bit cleaner and more complete than the ones from my childhood, but the lines and curves are much the same. I bought it as a boon to the memory of my younger self who tangled her legs in the vetch and limestone in search of natural treasures. I think she'd approve.

Appendix I: Further Resources

Vulture Culture 101 is meant to be an introduction to the fandom around dead things. There are plenty more books on taxidermy, tanning and related topics; I've listed a few here if you'd like to do more research into them. I recommend looking up reviews on Amazon and Goodreads to determine which ones will be the best for you to look into. Check out the bibliography, too!

How-To Books

Stuffed Animals: A Modern Guide to Taxidermy by Divya Anantharaman and Katie Innamorato

The Complete Guide to Small Animal Taxidermy: How to Work With Squirrels, Varmints and Small Predators by Todd Triplett

Home Book of Taxidermy and Tanning: The Amateur's Primer on Mounting Fish, Birds, Animals, Trophies by Gerald J. Grantz

Deerskins into Buckskins: How to Tan with Brains, Soap or Eggs by Matt Richards

The Ultimate Guide to Skinning and Tanning: A Complete Guide to Working with Pelts, Fur, and Leather by Monte Burch

Cultural and Artistic Explorations

Taxidermy Art: A Rogue's Guide to the Work, the Culture, and How to Do It Yourself by Robert Marbury

Taxidermy by Alexis Turner

Still Life: Adventures in Taxidermy by Melissa Milgrom

Curators: Behind the Scenes of Natural History Museums by Lance Grande

Humor

Crap Taxidermy by Kat Su

Also, just as a quick reference, my online database of animal part-related laws which I alluded to a few times throughout the book is at http://www.thegreenwolf.com/animal-parts-laws.

Appendix II: Recommended Nature-based Nonprofits

There are a lot of worthy environmental nonprofits out there. The following are some of my favorites to donate to. Keep in mind these are large, national and international organizations. You likely also have more local groups doing a lot of boots-on-the-ground work for your bioregion, too, so consider helping them with a donation and/or volunteer time.

The Xerces Society for Invertebrate Conservation
628 NE Broadway Ste. 200
Portland OR, 97232
USA
(503) 232-6639
http://www.xerces.org

When we think of protecting wildlife we often forget about insects and other invertebrates, yet these are the backbone of our ecosystems. The Xerces Society has done an incredible amount of work to both highlight the threats these animals face, and find solutions for their conservation.

The Nature Conservancy
4245 North Fairfax Drive, Suite 100
Arlington, VA 22203-1606
USA
(800) 628-6860
http://www.nature.org

Focuses on protecting habitats around the world, and educating people about the importance of healthy ecosystems. This includes direct protection of individual habitats in conjunction with local communities.

The Ocean Conservancy
1300 19th Street, NW
8th Floor
Washington, DC 2003
USA
800-519-1541
http://www.oceanconservancy.org

Works to protect the world's oceans and to create awareness of how crucial the oceans and their inhabitants are to the planet's health as a whole.

The Sierra Club
85 Second Street, 2nd Floor
San Francisco, CA 94105
USA
Phone: 415-977-5500
http://www.sierraclub.org

One of the oldest and largest environmental nonprofits, combines government lobbying with grassroots organization for a variety of ecological causes.
Natural Resources Defense Council
40 West 20th Street
New York, NY 10011
USA
(212) 727-2700
http://www.nrdc.org

Lobbies for the protection of both wild species and their environments, and is also instrumental in helping communities become more sustainable.

The Wilderness Society
1615 M St., NW
Washington, D.C 20036
USA
1-800-THE-WILD
http://www.wilderness.org

Many animals, plants and fungi that face extinction are vulnerable due to habitat loss; this group works to preserve wilderness areas, to include crucial habitat.

Appendix III: About the Writers

Lupa is an author, artist and naturalist living in the Pacific Northwest. She has been creating art from hides, bones and other dead stuff since 1998. She is the author of several books on nature and spirituality, as well as *The Tarot of Bones* deck and book. When she's not immersed in creative work she enjoys volunteering with local conservation organizations, taking care of a farm full of critters, and going hiking with her dog. More about Lupa and her works can be found at http://www.thegreenwolf.com.

Guest Essayists

A unique combination of science and style, **Divya Anantharaman** is New York City's premiere taxidermist. An award winning licensed professional preserving animals with honor and compassion, she took second place in the Professional division at the 2017 World Taxidermy Championships, received a Best in Show at the 2018 New England Taxidermy Championships, and is the coauthor of the book "Stuffed Animals: A Modern Guide to Taxidermy." Her specialties are birds, small mammals, and anatomic anomalies. She left the world of corporate fashion to pursue her love of natural history as the resident taxidermist at the Morbid Anatomy Museum, and has since built a following creating taxidermy, skeletal, and entomology displays, along with jewelry and wearable art, for anyone who finds wonder in the natural world. She works part time as a restoration assistant at the esteemed studio of Wildlife Preservations where her love of museum taxidermy grows through helping preserve historic treasures. She is a board member of the New England Association of Taxidermists. Her clients include museums, art galleries, and private collectors, though she really enjoys demystifying taxidermy for newbies through workshops and lectures. In her commitment to conservation, she regularly works with various organizations like the Audubon Society, and volunteers as a rescuer/rehabber. All animal parts are legally and sustainably obtained. Website gothamtaxidermy.com
IG @gotham_taxidermy

In 2011, **Ashley Cheeks** (better known as "blackbackedjackal") began her journey as an oddities collector, with an interest in canid pathologies and color mutations. Over the years, she progressed from simply collecting furs and bones to processing her own findings. By 2015, she honed her skinning and tanning skills to near professional levels, where even the worst roadkill could be saved. She takes interest in all forms of animal processing and is dedicated to learning any and all preservation methods. She believes that all remains that come into her care are worth salvaging, and works diligently to create artistic memorials to immortalize the creatures she works on.

Eric Foote is a professional tanner who runs his small tannery, Valley Fur Shed, out of his home in Northern California. From early childhood his interests have centered around science, art, and animals, which naturally led him to Vulture Culture when he was seventeen years old. In order to save a few bucks while starting his collection, he decided to learn how to tan furs on his own. Immediately he fell in love with the process of preserving animal remains as well as the opportunity it gave him to see in person and handle many various species he otherwise never could. Seven years and several hundred pelts later, Eric remains a proud Vulture with an ever-growing passion for the art of tanning and the animals he preserves."

Shelby Hendershot is a full-time purveyor of all manner of dead things. She has been reclaiming, restoring and respiriting natural products at her tannery in rural Okanogan County, Washington for more than 12 years. When not up to her elbows in her work, she is taking care of her cattle and goats.

Escher Null is an artist, former mortician, and Druidic Animist who has been working with taxidermy for over a decade. Their work has appeared in several past issues of the poetry and prose collection "A Sharp Piece of Awesome", and their preservation work has been displayed in the Arizona Museum of Natural History. They seek to preserve the spark inherent in all life, while also educating others on how to protect wildlife and the natural world.

Amy Wilkinson: I have always been interested in animal anatomy since I was a kid, but since finding the taxidermy community online back in 2013, and then the growth of the Vulture Culture community on Tumblr, I have allowed myself to really explore and experience things myself. It's especially heartening to bring a second life to an animal which has died, and allowing its remains to be further appreciated for its own unique sort of beauty.

Glossary

While not all of these terms may have shown up in the main body of this book, they're used commonly in Vulture Culture. You may see some *italicized* words within the definitions of other words which are also good to know; not all of them have their own independent entries in this glossary, and their meanings should be inferred from their use in the definitions of those other words. For example, in the definition for *Antler* below, you will also see the word *shed*, which does not have its own entry, but whose meaning is briefly described in the entry for *Antler*.

Antler: in certain members of the deer (cervid/*Cervidae*) family, these are horn-like bone structures that grow out of the skull used for defense and, in males, fighting for mates. Most species shed them yearly and grow new ones. The bone lives and is fed by blood vessels in the velvet skin that covers it while the antlers are growing, but the bone dies and the velvet and vessels are shed once growth is complete. An antler that falls off naturally is known as a *shed*.

Articulation: the practice of arranging a complete skeleton or part of a skeleton so as to resemble its shape when the animal was alive. Articulation generally involves attaching the bones to each other with wire and metal hardware, and occasionally other supports such as plexiglass rods and stands.

Beetle-cleaned: Bones that have had their flesh removed by dermestid beetles in the family *Coleoptera*, usually *Dermestes maculata*. These beetles eat decaying flesh and leave the bones intact, other than very small, delicate bones that may need to be removed before the beetles devour them.

Bleaching: See *Whitening*

Blown (Egg): an egg whose insides have been removed by creating a tiny hole in each end, and then blowing into one hole to force the contents out through the other.

Brain Tanning: a tanning process using the animal's brain, which has a substance known as *lecithin* that help preserve the hide by taking the place of the animal's oils that would normally cause the hide to rot. Brain tanning may also include a step in which the hide is cured in the smoke of a fire.

Cape: a hide used in taxidermy that consists of the head, neck, and shoulders/chest of the animal. The deer mounts you frequently see on walls are cape mounts.

Carcass: the remains of a deceased animal, particularly one whose flesh and skeleton are still more or less intact, though the skin may be removed.

Case Skinned: a hide that is not cut open on the back (dorsal) or belly (ventral), but is only open at the mouth and hind end. Also called a barrel skin because of its resemblance to a wooden barrel with the ends removed.

Conservation: the practice of protecting wildlife, plants, fungi and the environments they rely on. In contrast to many *environmentalists*, *conservationists* may support using the land and its living creatures for human consumption, such as hunting, fishing or logging. Contrary to many corporations involved in logging, mining and other activities, conservationists believe in protecting the land and its denizens from over-exploitation, excessive habitat loss, and extinction.

Craft-quality: a specimen that is damaged or otherwise primarily suitable as fodder for art projects. This may be further broken down into A, B, and C craft-quality hides, in which A is the best and C is the worst.

Cranium: the back portion of the skull which houses the brain.

Cruelty-free: see *Ethically Sourced*

Degreasing: the process of removing fat (grease) from bones that have been cleaned by maceration, dermestid beetles or other means. This generally involves immersing the bones in dish soap and water, sometimes with a heating element to speed up the process. A riskier process involves soaking in acetone without heat. Degreasing may take weeks or even months to complete.

Display-quality: a specimen that is in very good condition, though it may have minor damage.

Dorsal: refers to the back of an animal, as in a fish's dorsal fin. Hides that have been cut open down the back are referred to as dorsal-cut.

Dry Preserved: also, "dried." Refers to specimens that have been preserved through dehydration or dessication, such as bird wings that have been buried in salt or Borax to draw out moisture. Also see *Mummification* and *Rawhide*.

Dry Tan: a tanned hide that is allowed to dry out entirely from its immersion in water and other fluids. It must be rehydrated if it is to be used for taxidermy.

European Mount: a skull, most often of an animal with antlers or horns, in which the bone is completely cleaned of flesh, degreased and usually whitened, and then displayed on a wooden plaque.

Ethically Sourced: this is a contentious concept with no single agreed-upon definition. Some people consider only the remains of animals that died a natural death to be ethically sourced, while others include animals killed for food, as two possible definitions. While most Vultures are as honest about their specimens' origins as possible, a minority will claim anything is "ethically sourced" as a way to avoid criticism from Anti-Vultures, or to gain sales from people seeking to purchase *natural death* specimens.

Farmed: any animal that was raised by humans for a particular utilitarian purpose, whether for food or hides and other remains, or a combination thereof. Many Vultures are comfortable having the remains of cattle and other livestock as byproducts of food production, but may be less comfortable with fur from foxes and other wild species raised in confinement primarily for hides.

Fleshing: the process of removing muscle, connective tissues and other flesh from a hide before beginning preservation such as tanning.

Foramen Magnum: the large hole at the base of the skull where the neck attaches and the spinal cord connects to the brain.

Form: a polyurethane or, less commonly, wooden structure shaped like part or all of an animal's hairless body, with an emphasis on lifelike and correct muscle structure. *Taxidermy-quality* hides are stretched over these forms as a replacement for the animals' original flesh and bone.

Frozen: a specimen that is literally frozen to prevent decay, usually a hide, skull, or even entire *carcass*. It is kept frozen until it can be processed through tanning, bone cleaning or other appropriate methods. Skulls and bones may be frozen indefinitely without harm; however, a hide that has been frozen for more than a few months may lose hair, or *slip*, during the tanning process.

Fur: a skin with its hair still intact, whether *raw* or *tanned*. Fur is most often used to describe the hides of *furbearers*, such as foxes, coyotes, beavers and chinchillas, while the *hair-on* skins of *ungulates* like cows and deer is more often known as a *hide*. Pelt is used in much the same way as fur.

Green: see *Raw*

Hair Side: the side of a hair-on hide that has hair on it; this was the outside of the animal in life. Also see *skin side*.

Headdress: a wearable *tanned* animal hide, often for spiritual, cultural or costuming reasons. The head of the headdress may be reshaped to look more lifelike, with or without *taxidermy* materials like a *form* used in the process. It also has straps to hold the hide onto the wearer's head and sometimes body.

Hide: also known as a *skin*. This can refer to a *hair-on fur* or *pelt*, but can also refer to *rawhide* or *leather*. It is the skin of an animal with the flesh, bones and other parts removed. See also *raw*, *rawhide* and *tanning*.

Horn: in some ungulates such as cows, goats and antelopes, structures made of living bone and keratin *sheaths* that grow out of the skull used for defense and, in males, fighting for mates. Unlike antlers, horns are not shed but continue to grow throughout the animal's lifetime, with the sole exception of pronghorns which do shed the sheath yearly. The keratin sheath is made of the same substance as our hair and fingernails. One exception is the rhinoceros, whose horn is made entirely of keratin without a bone core.

Hunted: a wild or feral animal that has been pursued and killed by a human, usually with a gun or knife, though smaller animals such as frogs may be killed with spear-like objects. Some Vultures choose not to possess hunted remains, especially if they don't know whether the hunt was legal, or if the species is endangered.

Invertebrate: any animal without an internal backbone or spinal cord. Commonly seen invertebrates in Vulture Culture include insects, spiders, and other "bugs", as well as octopi, shellfish and coral. The processes for preserving invertebrates often vary greatly from those for preserving vertebrates.

Leather: any hide that has been *tanned* without the hair. Leather may be left natural, or given a coating of wax, plastic or a similar water-resistant substance, with patent leather being a popular example.

Lecithin: a substance found in the fat of animals and some plants that can be used to replace skin oils n hides that are being tanned. The lecithin in brains is most commonly used, though egg and soy tanning also rely on the lecithin of those substances.

Maceration: the process of cleaning the flesh from bones by allowing them to rot in a vat of water for weeks or months.

Mount: a piece of taxidermy created by stretching a hide over a form to make it look as lifelike as possible. A hard mount uses a foam (or less often, wood) form so that the entire piece is rigid. A soft mount involves a whole fur hide in which the head is mounted in the same way as a hard mount, but the rest of the body is stuffed with polyfill or another soft substance, occasionally with wires in the legs and tail to make them poseable.

Mountable: see *Taxidermy-Quality*

Mummification: a specialized dry preservation process in which the entire carcass is dehydrated to the point of little to no decay, generally without removing any hide or flesh. This may be done naturally, as in an animal that dies in the desert and dries out, or through specialized techniques.

Museum-quality: a specimen that is in the best possible condition, both in being free of damage and in being professionally prepared.

Mutation: refers to a coloration in a hide, usually a fox pelt, which is derived from a genetic mutation. Almost always used when speaking of unique colorations only found in farmed red foxes, as opposed to natural mutations like albinism or melanism (silver fox).

Natural Death: an animal that died of disease, injury, old age, or any other death not caused directly by humans or out technology. *Roadkill* is generally not considered a natural death.

Natural History: the scientific study of nature, particularly animals, plants and fungi, though it can also include geology, hydrology and other elements of an ecosystem. Natural history is generally more focused on observations than experiments, and is not solely limited to academics or laboratories. A practitioner of natural history may be known as a *naturalist*.

Nature-cleaned: bones that have been cleaned by wild insects (rather than captive dermestid beetles), bacteria and/or the elements of nature like wind and water. They often show signs of their exposure to the outdoors, such as a cracked, rough surface and weaker bone.

Open Skinned: a hide that has been cut open down the neck, chest and belly so that it can be laid out flat. Also known as *abdominal-cut* or *ventral-cut*.

Pathological: any specimen with an unusual deformity or other mutation that may have had a negative effect on the animal while it was alive. This can include bones warped by tumors or badly healed breaks, wet specimens with growths or congenital deformities like naturally missing or extra body parts, and so forth.

Pickling: a process in preparing a hide for tanning in which the hide spends twenty-four to forty-eight hours soaking in a mixture that includes an acid, a salt and water. This removes non-structural proteins from the fibers of the hide and makes it easier for tanning solutions to bind to the hide. Some tanners consider pickling to be optional, while others refuse to skip this part of the process.

Process: to process a specimen means to do something to it that will preserve it longer than it would last in nature. This can be as simple as bringing a weathered bone inside out of the weather, or as complex as the chemical changes in hide tanning. Most processed specimens may last for many years assuming the process was done correctly and the specimen is cared for properly.

Raw: a hide or bone (usually a skull) that has not been preserved yet, though it may have received very basic *processing*, such as removing the skin from a skull or taking a hide off a *carcass* and *fleshing* it.

Rawhide: a hide that has been dried but not tanned and is stiffer than a tanned hide. While it is possible to dry some hides simply by stretching them out and letting them dry in the air and sun, more commonly they are dried chemically with an alkaline substance like ash. This process removes the hair from the hide, after which any remaining flesh needs to be removed or it will rot.

Remains: see *Specimen*

Roadkill: an animal that has been struck and killed by a car, truck or other vehicle. Animals killed by trains or boats may also be included in this definition even though there's no actual road included. Some Vultures consider roadkill to be ethically similar or equal to *natural death* specimens, as the death was not intentional even though it was caused by humans.

Rug: a hide that has been cut down the *ventral* side to be laid flat, and then *tanned*. Those with the heads intact usually have them *mounted* using *taxidermy* methods.

Rug Cut: See *Ventral*

Salting: to preserve a hide by removing extra flesh and then covering the hair side with salt. This is generally considered a temporary preservation until the hide can be tanned or turned into rawhide.

Skin Side: the smooth, hairless side of a hair-on hide. This was the inside of the animal's skin.

Skinning: the process of removing the skin of an animal from the rest of the *carcass*. In spite of questionable videos on the internet, skinning is done when the animal is already dead and therefore no longer moving, as it is a tricky process, particularly when trying to avoid damaging the hide.

Skull: the bones in the head of a vertebrate animal; the jawbone in particular is also known as the maxilla.

Skull Cap: The portion of a *skull* to which *antlers* or *horns* are attached. Some people choose to cut the rest of the skull off, and only display the antlers/horns on the skull cap, especially if display space is limited or the rest of the skull was damaged.

Slipping: when a hair-on pelt loses hair, usually due to improper preservation, or preserving a hide that was too old or had begun to rot. You will generally see bald patches on the hide, especially on places where the skin is thinner.

Soft Mount: see *Mount*

Species: a particular type of animal, plant, fungus or other life form that is differentiated from other species by both outward physical characteristics and genes. For example, the red fox is a different species than the Arctic fox or gray fox, though a red fox and a silver fox are the same species since the silver fox is just a red fox with more melanin making its fur darker.

Specimen: Often used for parts used in scientific research, but may be used to refer to any preserved part of an animal. *Dead things* is a mildly irreverent, tongue in cheek synonym.

Stretch: to stretch a hide over a board or other flat surface during defleshing and tanning so as to keep it from shrinking too much or drying unevenly, and to more easily remove flesh.

Stole: a sort of scarf made with one or more whole tanned animal hides, usually mink, marten or fox. Each hide often has a very rudimentary *form* and glass eyes in its face; it may or may not have its feet, legs and tail intact. Most will also have cotton or another stuffing inside the hide. *Vintage* stoles are popular sources of hides for those who enjoy visiting antique shops and flea markets.

Tanning: the process of preserving an animal hide by removing flesh, natural skin oils and moisture and/or changing its pH so as to discourage rot, both by discouraging bacteria and encouraging chemical reactions between the substances used in the tanning process and collagen or other fibers in the hide. Hides may be tanned with the hair on in mammals, or without hair as in the leather of various mammal, bird, reptile and fish species. Some of the active substances in tanning include *lecithin* and *tannins*.

Tannins: natural or synthetic substances that are used to preserve a hide through drying it, changing its pH, and making it more bacteria-resistant. Natural tannins may be found in certain tree leaves, most famously, among other plant materials.

Taxidermy: a process by which a tanned hide is made to appear more lifelike, usually by being made into a *mount* or *rug*. The term may also be used as a more general marker for specimens, especially as a hashtag online (#taxidermy) so that Vultures know that a particular post is relevant.

Taxidermy-quality: a tanned hide that is particularly suitable for creating *taxidermy*; *wet tanned* hides are generally preferred over *dry tanned*, and the parts of the hide intended for taxidermy must be exceptionally intact without missing pieces, even as small as eyelids and the interior of nostrils.

Trapped: an animal that has been caught and/or killed in a trap created or set by humans. Remains of these animals may be controversial as animal rights activists question how humane trapping is, while trappers try to educate people on their practices.

Ventral: the underside of an animal, such as its belly and chest. Hides that have been cut open on the ventral side are referred to as *ventral-cut* or *open skinned.*

Vertebrate: any animal that has an internal backbone and spinal cord. Most vertebrates have more extensive skeletons than just a backbone. With the exception of some fish like lampreys and sharks, whose skeletons are composed mostly of cartilage, vertebrate skeletons are made of bone.

Vintage: a term with no specific definition, it's best described as "something older" though the most common definition is any item twenty years or older. Something that is one hundred years or older is said to be an antique.

Vulture: a member of *Vulture Culture.* Also any of a number of large scavenging birds in the raptor family. You do not have to know any Vultures of either sort in order to be part of Vulture Culture; your appreciation of dead things is enough.

Vulture Culture: the respectful appreciation of hides, bone and other preserved animal remains, often for scientific or aesthetic purposes. Often abbreviated as VC. Common hashtag variants include #vultureculture, #vulture culture, and #VC. Also see the rest of this book for more detailed information.

Wallhanger: any *display quality* or *craft quality* hide; the terms stems from some Vultures' practice of displaying imperfect hides by hanging them by their muzzles from tacks or nails on the wall.

Wet Specimen: a specimen preserved in formaldehyde, alcohol or another preservative chemical. Generally the specimen is not skinned or otherwise taken apart, but put whole into a jar of the preservative and then sealed. While many wet specimens are made with an entire carcass, others may preserve only a head, single limb or other individual part.

Wet Tan: a tanned hide that is not allowed to dry out after the tanning process; it may be wrapped in plastic and then put in the freezer still wet until it is time to use it for taxidermy.

Whitening: the process of making a bone whiter in color than its natural coloration and any staining it may have. Contrary to the popular synonym, bleaching, chlorine bleach should never be used for whitening bones as it will cause it to deteriorate. The most common whitening agent used is hydrogen peroxide.

Bibliography

Websites

American Mountain Men (various). *Mountain Men and the Fur Trade: Sources of the Fur Trade in the Rocky Mountain West.* Retrieved from http://www.mtmen.org/

Bentley, Andy (2007). *Shipping and Handling of Natural History Wet Specimens Stored in Fluids as "Dangerous Goods" – Hazardous Materials.* Retrieved from https://www.nps.gov/museum/publications/conserveogram/11-13.pdf

Blitz, Matt (2015). *Why Taxidermy Is Being Revived for the 21st Century: A new generation of young practitioners is leading a resurgence in this centuries-old craft.* Retrieved from http://www.smithsonianmag.com/arts-culture/why-taxidermy-being-revived-21st-century-180955644/?no-ist

BLM (unknown). *Collecting On Public Lands.* Retrieved from https://www.blm.gov/sites/blm.gov/files/documents/files/collecting_on_publiclands.pdf

Brewer, Sarina (unknown). *Introduction.* Retrieved from http://www.sarina-brewer.com/introduction.html

CITES (2019). *Convention on International Trade in Endangered Species.* Retrieved from http://www.cites.org

Clay, Joanna (2016). *Women Are Killing It In Taxidermy.* Retrieved from https://www.marketplace.org/2016/07/01/business/taxidermy

Dickinson, Joy Wallace (2003). *Search For Family Roots Leads To Special Garden.* Retrieved from http://articles.orlandosentinel.com/2003-05-11/news/0305090468_1_aunt-aggie-grandison-maddox-family

Fur Council of Canada (unknown). *Origin Assured.* Retrieved from http://www.furcouncil.com/originassuredfur.aspx

Hines, Alice (2015). *The History of Faux Fur: For more than 100 years, the fine line between finks and minks has been blurred.* Retrieved from https://www.smithsonianmag.com/history/history-faux-fur-180953984/?no-ist

Hugo, Kristin (2015). *What is Vulture Culture?* Retrieved from https://www.youtube.com/watch?v=YaPvAizNGeA

Hugo, Kristin (2017). *Thousands of Bats Slaughtered Annually in Asia End Up on eBay and Etsy for Artsy Americans.* Retrieved from http://www.newsweek.com/thousands-bats-slaughtered-annually-asia-end-ebay-and-etsy-artsy-americans-681147

Huizhongwu (2018). *Peta [sic] still pushing the idea that animals are 'skinned alive' in China.*
Retrieved from http://shanghaiist.com/2013/08/27/peta_still_pushing_the_idea_that_animals_are_skinned_alive_in_china.php

Kiser, Margot (2014). *Burundi's Black Market Skull Trade: Has the tiny nation of Burundi become ground zero for a new global black-market trade in human remains?* Retrieved from http://www.thedailybeast.com/articles/2014/01/26/burundi-s-black-market-skull-trade.html

Lewis, Geoffrey D. (2019). *Museum: Cultural Institution*. Retrieved from https://www.britannica.com/topic/museum-cultural-institution

Lifebender (2013). *A Name for the Bone Collecting Community?* Retrieved from http://lifebender.tumblr.com/post/63306554883/a-name-for-the-bone-collecting-community

Maddicott, Alice (2011). *Aunt Aggie's Boneyard*. Retrieved from http://andtheghostssosilver.blogspot.com/2011/01/aunt-aggies-boneyard.html

NOAA (unknown). *Laws & Policies: Marine Mammal Protection Act*. Retrieved from https://www.fisheries.noaa.gov/topic/laws-policies#marine-mammal-protection-act

Noir Arts & Oddities (unknown). *Wet Specimen Info & Care Tips*. Retrieved from https://www.noirartsandoddities.com/wet-specimen-care/

Pandika, Melissa (2019). *John Edmonstone: The Freed Slave Who Inspired Charles Darwin*. Retrieved from https://www.ozy.com/flashback/john-edmonstone-the-freed-slave-who-inspired-charles-darwin/31600

ReD (2017). *The Bare Bones of Vulture Culture: Find Bones With Sleight of Mind*. Retrieved from https://www.mookychick.co.uk/how-to/interesting-hobbies/bare-bones-vulture-culture.php

Ritchison, Gary (unknown). *BIO 342: Comparative Vertebrate Anatomy: Lecture Notes 2 - Vertebrate Skeletal Systems*. Retrieved from http://people.eku.edu/ritchisong/342notes2.htm

Rogers, Elizabeth (unknown). *Gone Forever*. Retrieved from https://baynature.atavist.com/gone-forever

Semic, Sara (2016). *How Taxidermy Became Cool Again: A vanguard of young, female and mainly veggie taxidermists are taking up the reins*. Retrieved from https://www.elle.com/uk/life-and-culture/culture/articles/a32430/how-taxidermy-became-cool-again/

The Bone Room (unknown). *About Our Products*. Retrieved from http://www.boneroom.com/store/c83/About_Our_Products.html

USFW (2015). *National Eagle Repository: Conserving America's Future*. Retrieved from http://www.fws.gov/eaglerepository/USFW (2017). *Digest of Federal Resource Laws of Interest to the U.S. Fish and Wildlife Service: Migratory Bird Treaty Act of 1918*. Retrieved from https://www.fws.gov/laws/lawsdigest/migtrea.html

USFW (2018). *Eagle Permits*. Retrieved from https://www.fws.gov/midwest/eagle/permits/index.html.

USFW (2018). *Endangered Species Act: Overview*. Retrieved from http://www.fws.gov/endangered/laws-policies/

Voon, Claire (2014). *Meet the Ladies Who Turn Animal Corpses Into Art: Women Are Dominating the Rogue Taxidermy Scene*. Retrieved from https://www.vice.com/en_us/article/women-are-dominating-the-rogue-taxidermy-scene-666

A few more websites may be found as suggested reading in the footnotes throughout this book.

Books and Other Hardcopy Sources

Burns, Ken (director) (2009). *The National Parks: America's Best Idea* [Motion Picture]. United States: Public Broadcasting Service.

Copeland, Bill (1977, December). *Aunt Aggie's Boneyard: ...old Lake City landmark lost to progress.* Page 14-b?

Dunn, Amber C and Smith, Mark (2011). *The Coyote: Facts and Myths About Living with This Wild Canid.* Alabama Cooperative Extension System. ANR-1413

Hansen, Thor (2012). *Feathers: The Evolution of a Natural Miracle.* New York: Basic Books.

Marbury, Robert (2014). *Taxidermy Art: A Rogue's Guide to the Work, the Culture, and How To Do It Yourself.* New York: Artisan Books.

Morris, Pat and Ebenstein, Joanna (2014). *Walter Potter's Curious World of Taxidermy.* New York: Blue Rider Press.

Perkins, May Vinzant (1953?). *Aunt Aggie's Bone Yard: Historic Old Garden of Lake City, Florida.* Publisher unknown.

Sherrill, Tawny: "Fleas, Furs, and Fashions: *Zibellini* as Luxury Accessories of the Renaissance", in Robin Netherton and Gale R. Owen-Crocker, editors, *Medieval Clothing and Textiles*, Volume 2, p. 121-150

Su, Kat (2014). *Crap Taxidermy.* New York: Ten Speed Press.

Suderman, Steve (director) (2013). *To Make a Farm* [Motion Picture]. Canada: FilmBuff

Weschler, Lawrence (1995). *Mr. Wilson's Cabinet of Wonders: Pronged Ants, Horned Humans, Mice on Toast, and Other Marvels of Jurassic Technology.* New York: Vintage Books.

Made in the USA
Middletown, DE
14 October 2022